# DIRECT-TO GENETIC TESTING

## Summary of a Workshop

Mary Fraker and Anne-Marie Mazza, *Rapporteurs*

Committee on Science, Technology, and Law
Policy and Global Affairs

Board on Life Sciences
Division on Earth and Life Studies

Forum on Drug Discovery, Development, and Translation
Board on Health Sciences Policy

Roundtable on Translating Genomic-Based Research for Health
Board on Health Sciences Policy

National Cancer Policy Forum
Board on Health Care Services

INSTITUTE OF MEDICINE *AND*
NATIONAL RESEARCH COUNCIL
*OF THE NATIONAL ACADEMIES*

THE NATIONAL ACADEMIES PRESS
Washington, D.C.
**www.nap.edu**

**THE NATIONAL ACADEMIES PRESS** 500 Fifth Street, N.W. Washington, DC 20001

NOTICE: The project that is the subject of this report was approved by the Governing Board of the National Research Council, whose members are drawn from the councils of the National Academy of Sciences, the National Academy of Engineering, and the Institute of Medicine. The members of the committee responsible for the report were chosen for their special competences and with regard for appropriate balance.

International Standard Book Number-13: 0-978-0-309-16216-6
International Standard Book Number-10: 0-309-16216-5

Additional copies of this report are available from the National Academies Press, 500 Fifth Street, N.W., Lockbox 285, Washington, DC 20055; (800) 624-6242 or (202) 334-3313 (in the Washington metropolitan area); Internet: www.nap.edu.

Printed in the United States of America

# THE NATIONAL ACADEMIES
*Advisers to the Nation on Science, Engineering, and Medicine*

The **National Academy of Sciences** is a private, nonprofit, self-perpetuating society of distinguished scholars engaged in scientific and engineering research, dedicated to the furtherance of science and technology and to their use for the general welfare. Upon the authority of the charter granted to it by the Congress in 1863, the Academy has a mandate that requires it to advise the federal government on scientific and technical matters. Dr. Ralph J. Cicerone is president of the National Academy of Sciences.

The **National Academy of Engineering** was established in 1964, under the charter of the National Academy of Sciences, as a parallel organization of outstanding engineers. It is autonomous in its administration and in the selection of its members, sharing with the National Academy of Sciences the responsibility for advising the federal government. The National Academy of Engineering also sponsors engineering programs aimed at meeting national needs, encourages education and research, and recognizes the superior achievements of engineers. Dr. Charles M. Vest is president of the National Academy of Engineering.

The **Institute of Medicine** was established in 1970 by the National Academy of Sciences to secure the services of eminent members of appropriate professions in the examination of policy matters pertaining to the health of the public. The Institute acts under the responsibility given to the National Academy of Sciences by its congressional charter to be an adviser to the federal government and, upon its own initiative, to identify issues of medical care, research, and education. Dr. Harvey V. Fineberg is president of the Institute of Medicine.

The **National Research Council** was organized by the National Academy of Sciences in 1916 to associate the broad community of science and technology with the Academy's purposes of furthering knowledge and advising the federal government. Functioning in accordance with general policies determined by the Academy, the Council has become the principal operating agency of both the National Academy of Sciences and the National Academy of Engineering in providing services to the government, the public, and the scientific and engineering communities. The Council is administered jointly by both Academies and the Institute of Medicine. Dr. Ralph J. Cicerone and Dr. Charles M. Vest are chair and vice chair, respectively, of the National Research Council.

**www.national-academies.org**

# Preface

At a televised ceremony on May 31, 2007, Richard Gibbs, director of the Human Genome Sequencing Center at the Baylor College of Medicine, presented DNA pioneer James Watson with a copy of his entire genome sequence. Two years later, in January 2009, Harvard University professor Steven Pinker famously announced in the *New York Times* that he was allowing his sequenced genome—and entire medical history—to be posted online in the Personal Genome Project's public database; a database designed to contain genomes and traits of 100,000 people.

Today, scores of companies, primarily in the United States and Europe, are offering whole genome scanning services directly to the public. These companies are web-based services that offer individuals living anywhere the opportunity to learn about their own personal genetic makeup. It is important to understand that, in general, these services do not offer the whole genome sequencing provided to Watson and Pinker but rather the sequencing of selected regions of the genome to examine different sets of specific alleles that have been reported in the scientific literature to be associated with particular clinical phenotypes. The associations, reported as "probabilities," confer a small increase in risk of developing certain diseases, but it is difficult even for experts to interpret their significance. Notwithstanding these concerns, many individuals have sought these services, and many news outlets have reported on and covered "spitting" parties where individuals contribute their saliva for genotyping—thus adding to the popularity and novelty of these tests.

Although the companies and their services have become highly visible, some argue that there is insufficient regulatory oversight to provide adequate quality assurance to consumers who use these tests. The proliferation of these companies and the services they offer demonstrate a public appetite for this information and where the future of genetics may be headed; they also demonstrate the need for serious discussion about the regulatory environment, patient privacy, and other policy implications of direct-to-consumer (DTC) genetic testing.

Rapid advances in genetic research already have begun to transform clinical practice and our understanding of disease progression. Existing research has revealed a genetic basis or component for numerous diseases, including Parkinson's disease, Alzheimer's disease, diabetes, heart disease, and several forms of cancer. The availability of the human genome sequence and the HapMap, plummeting costs of high-throughput screening, and increasingly sophisticated computational analyses have led to an explosion of discoveries of linkages between patterns of genetic variation and disease susceptibility. This research is by no means a straight path toward better public health; almost all of the major diseases facing humankind today have a genetic component—usually involving many genes, not just one—as well as environmental and behavioral components. Improved knowledge of the genetic linkages, however, has the potential to change fundamentally the way health professionals and public health practitioners approach the prevention and treatment of disease. Realizing this potential will require greater sophistication in the interpretation of genetic tests, new training for physicians and other diagnosticians, and new approaches to communicating findings to the public. As this rapidly growing field matures, all of these questions require attention from a variety of perspectives.

To discuss some of the foregoing issues, several units of the National Academies held a workshop on August 31 and September 1, 2009, to bring together a still-developing community of professionals from a variety of relevant disciplines, to educate the public and policy-makers about this emerging field, and to identify issues for future study. The workshop brought together physicians, scientists, lawyers, ethicists, patient advocacy groups, industry representatives, and other stakeholders to identify issues associated with the increasing availability and use of DTC genetic testing and appropriate policy responses. The meeting featured several invited presentations and discussions on the many technical, legal, policy, and ethical questions that such DTC testing raises, including: (1) overview of the current state of knowledge and the future research trajectory; (2) shared genes and emerging issues in privacy;

(3) the regulatory framework; and (4) education of the public and the medical community.

This workshop summary has been prepared by the workshop rapporteurs as a factual summary of what occurred at the workshop. The statements made are those of the rapporteurs or individual workshop participants and do not necessarily represent the views of all workshop participants, the planners of the workshop, or the National Academies.

# Acknowledgments

We gratefully acknowledge the efforts of the workshop planning group. The planning group helped identify topics for the workshop, framed the focus for each session, and suggested speakers. We wish to thank the planning group co-chairs Frederick Anderson, Jr., *Partner, McKenna, Long & Aldridge, LLP*; Barbara Bierer, *Professor of Medicine, Harvard Medical School and Senior Vice President, Research, Brigham and Women's Hospital;* and members Joseph Fraumeni, Jr., *Director, Division of Cancer Epidemiology and Genetics, National Cancer Institute*; Patricia Ganz, *Professor of Health Services, School of Public Health and Professor of Medicine, David Geffen School of Medicine, University of California, Los Angeles*; Mikhail Gishizky, *Chief Scientific Officer, Entelos, Inc.*; Alberto Gutierrez, *Deputy Director for New Product Evaluation, U.S. Food and Drug Administration;* Kathy Hudson, *Director, Genetics and Public Policy Center, Johns Hopkins Berman Institute of Bioethics* (until July 31, 2009); Muin Khoury, *Director, National Office of Public Health Genomics, Coordinating Center for Health Promotion, Centers for Disease Control and Prevention*; David Korn, *Vice Provost for Research, Harvard University*; and Jonathan Moreno, *David and Lyn Silfen University Professor, Center for Bioethics, University of Pennsylvania Health System.*

We also acknowledge the contributions of the following individuals who made presentations at the workshop: Timothy Aitman, *Professor of Clinical and Molecular Genetics, Division of Clinical Sciences, Imperial College London*; Susannah Baruch, *Policy Director, Generations Ahead and Policy Analyst, Genetics and Public Policy Center, Johns Hopkins University*; K. David Becker, *Chief Scientific Officer, Pathway Genomics Corporation*; Andrea Ferreira-Gonzalez, *Professor of Pathology, Virginia Commonwealth University and Director, The Molecular Diagnostics Labo-*

*ratory, Virginia Commonwealth University Health System*; Harvey Fineberg, *President, Institute of Medicine*; Katrina Goddard, *Senior Investigator, Kaiser Permanente Center for Health Research*; Alan Guttmacher, *Acting Director, National Human Genome Research Institute, National Institutes of Health* (until November 3, 2009); Courtney Harper, *Acting Director of the Division of Chemistry and Toxicology Devices, Office of In Vitro Diagnostic Device Evaluation and Safety, Center for Devices and Radiological Health, U.S. Food and Drug Administration*; Gregory Kutz, *Managing Director, Forensic Audits and Special Investigations, Government Accountability Office*; Sandra Soo-Jin Lee, *Senior Research Scholar, Stanford Center for Biomedical Ethics*; Elissa Levin, *Director, Genetic Counseling Program, Navigenics, Inc.*; Joseph McInerney, *Executive Director, National Coalition for Health Professional Education in Genetics*; Kathryn Phillips, *Professor of Health Economics and Health Services Research, University of California, San Francisco*; Scott Woodward, *Director, Sorenson Molecular Genealogy Foundation.*

This report has been reviewed in draft form by individuals chosen for their diverse perspectives and technical expertise, in accordance with procedures approved by the National Academies' Report Review Committee. The purpose of this independent review is to provide candid and critical comments that will assist the institution in making its published report as sound as possible and to ensure that the report meets institutional standards for quality and objectivity. The review comments and draft manuscript remain confidential to protect the integrity of the process.

We wish to thank the following individuals for their review of this report: Timothy Aitman, Imperial College, London; Naomi Aronson, Blue Cross and Blue Shield Association Technology Evaluation Center; Henry Greely, Stanford University; Bernard Lo, University of California, San Francisco; Channing Robertson, Stanford University; and Fay Shamanski, College of American Pathologists.

Although the reviewers listed above have provided many constructive comments and suggestions, they were not asked to endorse the content of the report, nor did they see the final draft before its release. The review of this report was overseen by Enriqueta Bond, President Emeritus, Burroughs Wellcome Fund. Appointed by the National Academies, she was responsible for making certain that an independent examination of this report was carried out in accordance with institutional procedures and that all review comments were carefully considered. Responsibility for the final content of this report rests entirely with the authors and the institution.

# Contents

# Introduction

In late 2008, *Time* hailed "The Retail DNA Test" as the "Invention of the Year,"[1] and *Nature* declared 2008 to be "the year in which personal genomics goes mainstream."[2] Today, many pharmacies in the United States sell a home DNA paternity test—often located alongside the array of pregnancy tests; the sample collection kit costs about $20, with an additional $130 to be mailed into the testing lab with the samples.[3] As Dr. Alan E. Guttmacher, former director of the National Human Genome Research Institute, and now Director of the National Institute for Child Health and Human Development observed, "Welcome to the Genome Era."

Major advances in genomic technologies in the early 21st century have helped to increase dramatically the number of genes identified as playing a role in a variety of common disorders.[4] These advances,

---

[1] Hamilton, A, TIME's Best Inventions of 2008, *Time Magazine*, October 29, 2008. Available at: http://www.time.com/time/specials/packages/article/0,28804,1852747_1854493,00.html, Accessed: September 16, 2010.

[2] Yeager, A, The Year In Which…, *Nature*, 2008; 456 (7224): 844-845.

[3] In May 2010, a DTC genetic testing company postponed plans to begin selling a similar home DNA test through a chain of retail pharmacies in the United States, after the Food and Drug Administration (FDA) informed the company that the test kit appeared to be subject to FDA's medical device review, clearance, and approval processes. The test was intended to enable customers to find out: their likelihood of developing any of 23 conditions including leukemia, Alzheimer's and certain cancers; whether or not they carry susceptibility genes for any of 23 genetic conditions that are known to be partly genetic—including diabetes; and how they are likely to respond to 10 substances ranging from caffeine to certain drugs for treating high cholesterol and breast cancer. The FDA subsequently sent letters to other DTC genetic testing companies indicating that their products qualified as devices under FDA rules, and therefore required premarket approval.

[4] The subjects of the workshop and of this report are DTC genetic tests, which address both genetic diseases and genetic disorders—terms that are often, though not always or entirely, interchangeable. For simplicity, the report uses the term "genetic disorder" throughout—unless quoting or closely paraphrasing a statement by one of the presenters, or in cases where the term "disease" seems more appropriate.

### Single-gene and Multi-gene Disorders

It is important to distinguish between testing for single-gene (Mendelian) and multi-gene (complex) disorders. In Mendelian disorders, a "defective" gene may confer a virtually 100 percent probability for the manifestation of the disorder (Huntington's, for example) or a significant increase in probability (as is the case with the "breast cancer genes" BRCA1 and BRCA2).

Mendelian disorders (at least those that have been characterized to date) generally occur in relatively defined and often small populations, with a prevalence of only one in one-thousand—or fewer—of the total population; because of this, many of them are classified as "rare diseases." Testing for Mendelian disorders occurs primarily within the traditional medical setting, and test results indicate a specific increase in probability—information that is clinically useful.

In contrast, complex disorders tend to be quite common, including the "big four"—Alzheimer's, cancer, cardiovascular disease and diabetes—and in fact are sometimes referred to using the shorthand "common diseases." Tests for complex disorders are commercially available primarily through direct-to-consumer genetic testing companies. The frequency with which many complex disorders occur, combined with the very human desire to know—and the wish thereby to "control"—one's future, have increased demand for DTC genetic testing. Unlike genetic tests for Mendelian disorders, however, tests for complex disorders are merely predictive of an altered risk associated with developing a disease. However, with the current information on SNP associations, the absolute risks tend to be low and are not of significant utility to alter standard of care.

combined with significant reductions in the cost of genetic tests, have spawned a new business model in which companies market genetic tests and personalized genetic profiles directly to consumers.

The rapid growth and evolution of the direct-to-consumer (DTC) genetic testing, "defined as genetic tests and services that are advertised directly to consumers, that are purchased through consumer-initiated requests, and that provide test results directly to the consumer without the involvement of the consumer's health care provider," has increased the focus on a number of issues raised by genetic testing.[5] These include scientific, clinical, and technical issues; economic and financial issues;

---

[5] *Direct-to-Consumer Genetic Testing*, Report of the Secretary's Advisory Committee on Genetics, Health, and Society, April 2010.

and free-market and first-amendment issues of honesty, fidelity to facts and truth-in-advertising—as well as the centuries-old tension between individual rights and authority, regulation and social responsibility.[6]

Although many of these issues also arise in the context of genetic testing in the traditional clinical setting, the rapid and continuing growth of the DTC industry—estimated to be approaching a billion dollars per year—has increased the urgency to address them. To further both the discussion of these issues in the DTC context and to identify approaches for addressing them appropriately, the National Academies convened a multi-disciplinary panel of experts for a workshop on DTC genetic testing, held in Washington, DC, on August 31 and September 1, 2009.

The workshop considered a number of questions relating to DTC genetic testing including:

- *Clinical value.* Beyond the strictly technical criteria of a test's sensitivity, specificity, and analytical validity, what are the associated positive and negative predictive values, and how useful to the consumer is the resulting information? Can the results be used to predict disease susceptibility, and do they provide enough information to enable informed decision making about preventive measures and treatment, including lifestyle changes, that can affect patient outcomes?
- *Commercial, financial and economic impact.* What are the tests' potential effects on the economy, including changes in GDP? Will they produce savings to individuals and the health care system, or will they increase costs?
- *Social value.* Will the tests create a better-informed society and encourage positive lifestyle changes? Will they facilitate the formation of useful online patient communities and support groups?
- *Role in medicine and medical research, and appropriate regulation.* Are DTC genetic testing companies practicing medicine? Do they have a therapeutic relationship with their customers? If so, what are the regulatory considerations that come into play? When companies use customers' data for research, do those individuals, become *de facto* "human research subjects?" If so, are they adequately protected? Are the companies compliant with applicable federal regulations, and what oversight mechanisms need to be in place? As large-scale amassers of genetic data, what are the companies' responsibilities to provide access to those data by researchers, customers, and society as a whole? Should

---

[6] From the remarks of **Harvey Fineberg**, President, Institute of Medicine.

DTC genetic testing companies be subject to the same legal, regulatory, and ethical regimes as are clinicians and diagnostic and medical research laboratories?

• *Risks to individuals, the health care system, and society.* In addition to potential financial and economic risks, what are the risks of overuse, underuse, or inappropriate use of these tests and the resulting information? What are the privacy issues relating to customers and their extended kinships? Should the customers own and control their genetic samples and data? What measures can be taken to ensure that customers are giving fully informed consent? How can public education be improved regarding the risks and benefits of genetic testing and the role of genetic counselors?

The organizers recognized that this array of issues and questions could never be fully addressed during the course of a one and a half day workshop—especially given the deeply personal nature of genetic testing, the dynamism of the DTC industry and marketplace, and the profound implications, for both individuals and society as a whole, of the rapidly growing industry and its potential mechanisms of oversight. It is hoped, however, that the workshop and this report will lead to further study, discussion and, ultimately, a way forward.

This report summarizes the presentations and discussions at the workshop, organized in the following way: (1) the scientific and ethical foundations for DTC genetic testing—genomic associations, analytic and clinical validity, and clinical utility; (2) personal and social issues arising with regard to DTC genetic testing, particularly privacy, public awareness and understanding, and the need for broad public education; (3) research and medical issues in the context of DTC genetic testing—notably informed consent, the role of institutional review boards, and whether or not the customers of DTC genetic testing companies should be considered human research subjects; (4) the impact on health care and public health, looking particularly at economics, physician awareness and education, and genetic counseling; and (5) an overview of the current legislative and regulatory framework governing genetic testing in the United States—particularly the Genetic Information Nondiscrimination Act (GINA), the Clinical Laboratory Improvement Amendments (CLIA), and the Health Insurance Portability and Accountability Act (HIPAA). Finally, at the conclusion of each section, there is a series of "Questions Raised for Further Discussion" that relate to the content of the specific section. The report concludes with a holistic examination of a number of areas for further study. Appendixes include a glossary of terms and acronyms, an overview of how DTC genetic testing is currently regulated outside the United States, a list of currently available DTC genetic tests, and a representative list of DTC genetic testing companies. Additional appendixes provide biographies of the workshop's planning group, the workshop agenda and participants, and the membership of the oversight boards and committees.

In many places, this report summarizes the commentary of workshop participants. In such cases, speaker attribution is provided in the report's footnotes.

# Scientific Foundations for Direct-to-Consumer Genetic Testing

More than a century before the human genome sequence was fully revealed in 2003, biomedical researchers had begun to recognize that certain human diseases were hereditary and carried a very high probability of being transmitted to offspring; that is, such diseases behaved as Mendelian disorders. Over the decade preceding the completion of sequencing the human genome, researchers were beginning to identify specific genes for some of these Mendelian disorders: for example, cystic fibrosis in 1989, Huntington's in 1993, and breast cancer in 1994 and 1995 (BRCA1 and BRCA2, respectively). And this was only the beginning.

## GENOMIC ASSOCIATIONS[1]

By 2002, genes had been identified for roughly 1,700 of the 5,000-6,000 known Mendelian human disorders, most of them resulting each from a single erroneous allele, or version of a gene; in contrast, fewer than 10 genes associated with more common, complex human disorders had been discovered.[2] It appeared that with then-current technologies, the cost to locate all the single-nucleotide polymorphisms (SNPs)—individual variations in the coding DNA sequence of the genome—associated with any one complex disorder would be about $10 billion: far too expensive to employ in a full-out assault on common diseases.

---

[1] Unless otherwise noted, the section "Genomic Associations" reflects the remarks of **Alan Guttmacher**, former director of the National Human Genome Research Institute, National Institutes of Health.

[2] Glazier A, Nadeau J, Aitman T., Finding genes that underlie complex traits, Science, 2002: 298 (5602): 2345-2349.

Research conducted during the early 2000s revealed that sets of SNPs within distinct regions of the chromosome tend to be inherited together. If specific SNPs were contributing to common disease progression, identifying variants between distinct populations could theoretically help construct a map of the genome that identifies areas of potential association with disease. These initial observations led to the formation of an international consortium called the Haplotype Map (HapMap) project, which resulted in the creation of a HapMap, a catalogue of SNPs commonly found in a consensus human genome and their locations within the genome. Using the HapMap, a researcher can conduct a genome-wide association study (GWAS)—a rapid scan of many individuals' entire genomes to discover alleles associated with a particular disorder. DNA microarray chips—miniature glass chips encoded with thousands of short, synthetic, single-stranded DNA sequences representing identified SNPs—have been developed to facilitate this approach. The technology relies on the discriminatory nature of DNA sequences to bind to exactly complementary pieces of DNA with high affinity. Genomic DNA, for example, can be denatured into single strands, cut into smaller pieces, labeled fluorescently, and allowed to bind to the DNA encoded on the chip. Differences between the binding pattern of DNA from individuals with a particular disease and the binding pattern of DNA from a control group can lead to the identification of specific alleles which may contribute to disease progression. A GWAS is particularly useful for finding alleles that contribute to common diseases, and it enables researchers to do so far more easily and cheaply than before. However, as sequencing costs continue to fall, many predict that the "$1,000 genome" will soon become a reality. At this price, it is conceivable that whole genome sequencing will eventually replace genome wide association studies.

Although a GWAS can be a powerful source of information about genetic predisposition to disease, so far these studies explain only a very small fraction of heritability and fail to capture other contributing factors such as multiple common genetic variants acting together, copy number variants, or epigenetics—chemical changes in base pairs, or physical changes in chromosome structure—that may greatly modulate phenotypic expression but that are themselves typically not heritable.

Therefore, a GWAS is a weak forecaster of an individual's risk for a genetic disorder. Obesity linkages, for instance, account for less than 2 percent of variance in heritable body mass index. Furthermore, a GWAS has a major drawback: it misses rare genetic variants that, when present, may have major health effects. As a result, assessing the value of DTC genetic tests only begins with determining whether or not they are scientifically valid—a fairly straightforward task. Much more problematic is determining whether or not they are clinically useful.

## ANALYTIC AND CLINICAL VALIDITY AND CLINICAL UTILITY[3]

As with any biological test, assessing the analytic validity of a particular genetic test is relatively straightforward:  does the test correctly measure or detect what it is intended to measure or detect? Specifically:

- **Analytic sensitivity:** How often a test result is positive when the genetic variant of interest is present in the tested sample.
- **Analytic specificity:** How often a test result is negative when the tested sample does not contain the genetic variant of interest.

Documenting analytic validity is important even though it may not be publicly disclosed.  Many of the tests currently offered by reputable DTC genetic testing companies have been assessed for analytic validity, sensitivity and specificity (although there is no current regulatory scheme that assesses or ensures this). A recent exception came to light just prior to the workshop: a software glitch had caused one DTC genetic testing company to confuse human and animal mitochondrial DNA.[4]

Assessing other factors in clinical validity, however—particularly a test's value in predicting disease and calculating risk—presents more complex challenges, many of them in the context of GWAS SNPs:

- **Credible genetic associations:** Whether or not a clear association has been established between a particular SNP and an increased likelihood of developing a specific disease.
- **Positive predictive value:** The probability that—if a target SNP is present in a tested subject's genome—the individual will eventually be affected by the associated disorder.

---

[3] Unless otherwise noted, this section reflects the remarks of **Muin J. Khoury**, director of the Centers for Disease Control and Prevention's (CDC) Office of Public Health Genomics and a National Cancer Institute Senior Consultant in Public Health Genomics.

[4] Aldhous, P., My "non-human" DNA: a cautionary tale, *New Scientist*, August 26, 2009. Available at: www.newscientist.com/article/dn17683-my-nonhuman-dna-a-cautionary-tale.html; Accessed: May 10, 2010.

• **Negative predictive value:** Similarly, the probability that—if a target SNP is not present in a tested subject's genome—the individual can be confident of never being affected by the associated disease.

• **Additional predictive value:** Whether or not a test provides predictive information beyond what could be readily learned by other means, such as a family history, clinical evaluation, or standard laboratory testing. Looking at heart disease, for example, it takes an estimated 20 or more SNPs to get a two-fold increased risk, whereas family history has much greater predictive value; it could be therefore, that awareness of SNPs contributes little or no additional value. In at least two cases, however—coronary heart disease (nine SNPs) and breast cancer (seven SNPs)—it appears that awareness of SNPs may indeed contribute appreciable value by stimulating earlier screening for these diseases.[5] The utility of this information for predicting individual risk, however, remains to be determined.

• **The uncertainty of risk estimation.** In addition to the problem of hidden heritability (the possibility of as-yet undiscovered risk alleles), there are factors other than one's genome—one's "raw DNA"—that can significantly influence the probability that an individual will eventually be affected by a genetic disorder. These factors include environmental and behavioral factors, epigenetics, the epidemiology of the disease, and the individual's age at testing.

Even more problematic than assessing clinical validity is the question of clinical utility. Assessing clinical utility—the net balance of benefits and harms to the individual, the population at large, and the overall health care system—is a more complex, and often more subjective, task. Much of the discussion of clinical utility revolves around actionability. Regarding genetic disorders for which no prevention or intervention is currently known, the practical value of testing for increased risk is often unclear. For a few devastating diseases—Huntington's and Alzheimer's for example—it is understandable that an individual might want to know in order to make financial or other arrangements. Unlike Huntington's disease, for which a positive test carries a virtually 100 percent probability of developing the disease, risk assessment for Alzheimer's disease is more complex and problematic and far less certain.

In addition, preventive actions that might be taken in response to increased risk for common diseases such as diabetes, cancer and cardio-

_____

[5] This sentence taken from the remarks of **Timothy Aitman**, Professor of Clinical and Molecular Genetics, Division of Clinical Sciences, Imperial College of London.

vascular disease—smoking cessation, exercise, a healthier diet—are not unlike the advice most individuals already receive from their health care providers, especially in the presence of relevant family histories or clinical risk factors. Moreover, a negative result might cause an individual to forego prudent prevention and early-detection measures. Furthermore, it is at least plausible that the kinds and efficacy of patient/consumer interventions in response to this information are affected by the extent of counseling or other professional intermediation in providing the information.

In December 2008, the NIH and CDC jointly sponsored a multidisciplinary workshop, "Personal Genomics: Establishing the Scientific Foundation for Using Personal Genome Profiles for Risk Assessment, Health Promotion, and Disease Prevention" in Bethesda, Maryland. The workshop report[6] made several recommendations:

- **Develop and implement industry-wide scientific standards.**
- **Develop and apply a multidisciplinary research agenda**, including observational studies and clinical trials, to fill knowledge gaps in the clinical validity and utility of personal genomics.
- **Enhance credible knowledge synthesis and dissemination** of information to providers and consumers.
- **Link scientific research on validity and utility to evidence-based recommendations** for use of personal genomic tests.
- **Consider the value of personal utility.**

One final point, which is neither directly nor exclusively relevant to genetic testing, *per se*, is nevertheless sufficiently important to raise in this context. One long-lasting and far-reaching benefit of the Human Genome Project to scientific research—and one that is often overlooked in favor of headline-grabbing advances in genomics—is the landmark agreement among participating centers to post their sequencing data every 24 hours. As beneficial as this advance has been in many ways, it has highlighted the need to balance three imperatives: enabling rapid public access to all data generated from publicly funded research; encouraging research by offering researchers exclusive use of their data for a period of time; and protecting participants' rights, including privacy. As a result, the National Institutes of Health is in the process of developing trans-NIH principles on data access.[7]

---

[6] The full report of the workshop was published in *Genetics in Medicine*, August 2009.

[7] This paragraph taken from the remarks of **Alan Guttmacher**, former director of the National Human Genome Research Institute, National Institutes of Health.

## QUESTIONS RAISED FOR FURTHER DISCUSSION

• What personal, familial, and societal issues are raised by DTC genetic testing and what issues will be raised upon the advent of affordable whole genome sequencing? How can individuals and society prepare for them?

• In contrast with fairly objective criteria of analytic and clinical validity, what should the criteria be for clinical utility, and who should make that determination—researchers, physicians, individuals, governments, society?

• Is personal utility a sufficient rationale for genetic testing?

• With the cost of locating all the SNPs associated with genetic contributions to any complex disorder rapidly becoming affordable, what process and criteria will be used to decide which common diseases will be explored and in what order?

• If DTC genetic testing companies are capable of confusing human and animal DNA, should they be providing genetic testing services to customers at all?

# Personal and Social Issues

Not surprisingly, DTC genetic testing raises a multitude of issues related to individuals' rights and responsibilities and to how genetic information can affect their relationships with family members and friends, as well as their overall social, emotional and financial well-being. A number of these issues are common to many areas of human health and medical care. Two issues with far-reaching implications are privacy and public awareness, understanding, and education.

## PRIVACY[1]

Privacy has been a long-standing issue in health care and medical research.[2] In December 2000, the Clinton administration released the Medical Information Privacy Regulation mandated by the Health Insurance Portability and Accountability Act of 1996 (HIPAA). The incoming Bush administration recalled the regulation, opened it to public review and comment, and in 2002 issued a revised regulation maintaining stringent protection of privacy for all covered personally identifiable health information, but easing access to such information for research and certain hospital administrative uses.

---

[1] Unless otherwise noted, the section "Privacy" reflects the remarks of **Susannah Baruch**, Policy Director, Generations Ahead, and Policy Analyst, Genetics and Public Policy Center, Johns Hopkins University.

[2] See for instance, the National Academies' publications *Privacy Issues in Biomedical and Clinical Research*, 1998 (Available at: www.books.nap.edu/catalog.php?record_id=6326); *Protecting Data Privacy in Health Services Research*, 2000 (Available at www.nap.edu/catalog.php?record_id=9952), and *Beyond the HIPAA Privacy Rule: Enhancing Privacy, Improving Health Through Research*, 2009 (Available at www.nap.edu/catalog.php?record_id=12458); Accessed May 12, 2010.

Genetic privacy in particular has become a matter of enormous concern to medical ethicists, policy-makers and the general public. The reasons include (1) a sense that one's own genome is a unique and ultra-private "personal diary of the future"; (2) arguments about fairness ("my genes are not my fault"); and (3) fear of discrimination and mistreatment based on genetic information.

Testifying in March 2007 before the U.S. House of Representatives' Energy and Commerce Committee's Subcommittee on Health, Francis Collins, then-director of the National Human Genome Research Institute, cautioned that "[F]ear of genetic discrimination threatens to slow both the advance of such groundbreaking biomedical research and the integration of the fruits of that research into our nation's health care. If individuals continue to worry that they will be denied health insurance or refused employment because they have a predisposition to a particular disease, they may forego genetic testing that could help guide medical professionals to lessen their risk."[3]

Such concerns had spurred the Human Genome Project in 1989 to designate three percent of its total budget for ongoing study of the project's ethical, legal and social implications (ELSI)—notably, though not exclusively, privacy and discrimination. Twenty years later, the Genetic Information Nondiscrimination Act (GINA) took effect in November 2009. In simplest terms, GINA protects individuals against access to and adverse use of their genetic information by health insurers and in the workplace.

GINA prohibits health insurers from requesting, requiring or using a person's genetic information in determining eligibility or in setting premium or contribution amounts. Although GINA does not prohibit insurers' use of genetic information in underwriting life, disability or long-term-care insurance, several states are considering such prohibitions (long-term-care insurance by 10 states, and life and disability insurance by 16 states each).

GINA also prohibits employers from requesting, requiring, purchasing or using genetic information about an individual or family member in any job-related decision. This suggests—though it is not explicitly

---

[3] Testimony by Francis Collins, Director National Human Genome Research Institute on "The Threat of Genetic Discrimination to the Promise of Personalized Medicine" before the U.S. House of Representatives Committee on Ways and Means' Subcommittee on Health, March 14, 2007. Available at: www.hhs.gov/asl/testify/2007/03/t20070314a.html, Accessed: May 12, 2010.

clear—that individuals who join online communities formed around genetic diseases are protected from any consequences of employers' intrusive Internet searching. If this is indeed the case, not only would this benefit individuals who join such networks seeking support, advice and information about their diseases, but it might also protect members of online communities formed around ethnic origin against employers' using ethnicity as a surrogate for health information (the increased risk of certain cancers among Ashkenazi Jews, for example). Moreover, the willingness of more members of genetic-disease and ethnic groups to join online communities without fear of reprisal would help researchers more rapidly identify relevant populations of statistically significant size.

What remains unclear at this point is whether GINA extends to genetic information that an employer might request as part of a voluntary workplace wellness program. Such information could range from family history to genetic testing offered through the workplace—a new market that some DTC genetic testing companies are looking to address.

GINA applies to the results of any and all genetic tests performed in the United States. HIPAA, however, does not: it only applies to statutorily defined "covered entities" and does not extend to DTC genetic testing companies. Consumers are unlikely to be aware of this technicality, making it crucial that a DTC genetic testing company's customers fully understand the company's privacy policies, including rules regarding the sale and disclosure of clients' data, use of the data in research, and—importantly—under what circumstances those rules can be changed.

DTC genetic testing companies generally define their privacy policies in terms of "appropriate" use of genetic information, and some describe their policies as complying with HIPAA. Without HIPAA's explicit privacy protections and the force of law, however, it is currently unclear what does and does not qualify as "appropriate." Neither is it clear what procedures are necessary and sufficient to protect the security and privacy of customers' samples and genetic information in the companies' possession.

Moreover, DTC genetic testing companies encourage customers to discuss test results with their physicians. As soon as such a discussion occurs—particularly if it results in medical advice, treatment or referral to a specialist—it becomes part of a patient's medical record, which can be requested by life, disability, and long-term-care insurers.

# PUBLIC AWARENESS, UNDERSTANDING, AND EDUCATION[4]

Although two national surveys, taken in 2006 and 2008, indicate a substantial increase in public awareness of DTC genetic tests, a majority of the national population continues to be unaware of them.

The 2006 survey,[5] conducted for the Kaiser Permanente Center for Health Research, found that higher incomes and more education correlated with greater awareness of DTC nutrigenomic genetic testing. Among respondents with low, intermediate and high family incomes, the percentages aware of the tests were 11, 13, and 16, respectively. More college graduates (19 percent) were aware of the testing than respondents with either limited or no college education (15 and 9 percent, respectively). Among the 14% of 2006 survey respondents who were aware of nutrigenomic DTC genetic testing, nearly three-quarters (73 percent) had heard or read about it via the media—either a magazine, a newspaper, or television. Far fewer (less than 10 percent of aware respondents) had learned about it from either the Internet or a health care provider, but this was in 2006—before most DTC genetic testing companies were actively marketing their products and services on the Internet.

Predictors of awareness in the more general 2008 survey,[6] funded by the CDC, included age, gender, education, race, and ethnicity. About one-third of one percent of respondents had used DTC genetic testing; of those tested, two-thirds had shared results with health care providers.

Participation in both surveys was high: the response rate in 2006 was 80 percent, with 5,250 participants; the rate in 2008 was 77 percent, with 5,399 participants.

In terms of information now available online, a study published in 2009[7] compared the information about genetic testing for venous throm-

---

[4] Unless otherwise noted, the section "Public Awareness, Understanding and Education" is taken from the remarks of **Katrina Goddard**, Senior Investigator, Kaiser Permanente Center for Health Research.

[5] Goddard K, Moore C, Ottman D, Szegda K, Bradley L, Khoury M., Awareness and use of direct-to-consumer nutrigenomic tests, United States, 2006, *Genetics in Medicine*, 2007, 9(8):510–517.

[6] Kolor K, Liu T, St Pierre J, Khoury M., Health care provider and consumer awareness, perceptions, and use of direct-to-consumer personal genomic tests, United States, 2008, *Genetics in Medicine*. 2009, 11(8):595.

[7] Goddard K, Robitaille J, Dowling N, Parrado A, Fishman J, Bradley L, Moore C, Khoury M., Health-related direct-to-consumer genetic tests: a public health assessment and analysis of practices related to Internet-based tests for risk of thrombosis. *Public Health Genomics*, 2009, 12 (2): 92-104.

boembolism, a circulatory disorder, found on the websites of five DTC genetic testing companies, the American College of Medical Genetics (ACMG) and the College of American Pathologists (CAP). The comparison found that the commercial sites tended to use simplified language, the effect of which was not only to make it more understandable for the general public, but also to lead the public to believe that testing is indicated for a broader class of persons. Moreover, in contrast with the ACMG and CAP sites, both of which listed several benefits and several risks of genetic testing, two of the commercial sites listed fewer of both, and the other three provided no information whatsoever on either benefits or risks.

Another study,[8] also published in 2009, looked at the impact of risk information on public attitudes toward DTC genetic testing. Women who received information about risks associated with BRCA testing were significantly less disposed toward being tested than those who received no information; they also had a markedly higher preference for genetic testing in a clinical setting over DTC testing.

A preceding study[9]—published in 2005—also using BRCA as an example, examined the effect of advertising on both the public's interest in a genetic test and also its understanding of the test's applications. After the advertising campaign, the number of women expressing interest in BRCA testing more than tripled. Comparing the women who expressed interest in BRCA testing before and after the

> "... the ability to receive test results ... off the grid, out of sight, out of pocket—to keep it under your control ... for people who care about privacy, [direct-to-consumer] may seem particularly appealing as an option. But ... if the information you get direct-to-consumer actually matters to your health, it's not going to stay private."
>
> Susannah Baruch
> Policy Director,
> Generations Ahead and
> Policy Analyst, Genetics and
> Public Policy Center,
> Johns Hopkins University

---

[8] Gray S, O'Grady C, Karp L, Smith D, Schwartz J, Hornik R, Armstrong K. Risk information exposure and direct-to-consumer genetic testing for BRCA mutations among women with a personal or family history of breast or ovarian cancer, *Cancer Epidemiology, Biomarkers & Prevention*, 2009; 18(4):1303-11.

[9] Mouchawar J, Hensley-Alford S, Laurion S, Ellis J, Kulchak-Rahm A, Finucane M, Meenan R, Axell L, Pollack R, Ritzwoller D., Impact of direct-to-consumer advertising for hereditary breast cancer testing on genetic services at a managed care organization: a naturally-occurring experiment. *Genetics in Medicine*, 2005, 7(3):191-197.

advertising campaign, more than two-thirds of those interested before the campaign were appropriate candidates for the testing; after the campaign, a bit less than half were appropriate candidates. The campaign clearly succeeded in generating interest in the test; it was far less successful, however, in clarifying who might and might not benefit from it and why.

As the results of genetic research increasingly inform health care and influence medical decisions, consumers will need to understand genetic tests' risks and benefits, their relevance and limitations. To encourage a common language that facilitates discussions between patients and their health care providers, education for the public and for health professionals should be complementary.[10]

## QUESTIONS RAISED FOR FURTHER DISCUSSION

• How can consumers' genetic privacy be protected, and how can consumers be reassured that their genetic information will indeed remain private?

• Should GINA explicitly extend to employers' wellness programs and to genetic information gathered on the Internet?

• Should HIPAA's privacy protections be extended explicitly to genetic information, no matter where it is collected and stored?

• Should there be a single privacy policy to which all DTC genetic testing companies are required to adhere? If so, how should it be developed and enforced?

• Should there be guidelines that all DTC genetic testing companies must follow in describing the risks, benefits and appropriate applications of the tests they offer? If so, how should these be developed and enforced?

• What is the most effective way of educating the general public on genetics, genetic testing, and risk?

• What kinds of interventions will be effective in helping to reduce disparities in awareness and understanding of genetics and genetic testing among ethnic and socioeconomic groups?

---

[10] From the remarks of **Joseph McInerney**, Executive Director, National Coalition for Health Professional Education in Genetics.

# Research and Medical Issues

The Internet is a tool that enables DTC genetic testing companies to reach and serve a large, diverse market and to do this almost entirely outside the institutions and constraints of medicine and medical research. In the process, these companies are rapidly amassing an unprecedented wealth of genetic samples and data, a resource that might be of great value in medical genetics research, although it must be recognized that the related phenotypic information proffered by clients is likely to be incomplete and of uneven quality.

Are DTC genetic testing companies actually practicing medicine? Do they have a therapeutic relationship with their customers—and the responsibilities such a relationship entails? At what point does a relationship become therapeutic? Does the employment of genetic counselors imply that a therapeutic relationship exists? DTC companies maintain that it is *not* their intent to practice medicine or to enter into a therapeutic relationship with their clients, and they do not consider themselves to be doing so.[1]

Are the companies' customers *de facto* human research subjects? What are customers' rights with respect to their samples, data, and any benefits of the research conducted using them? Should these companies' activities be governed by institutional review boards (IRBs) or other oversight bodies, such as HIPPA Privacy Boards or similar bodies addressing risks to privacy? Are their current informed consent documents and procedures sufficient to protect clients?

---

[1] This sentence reflects the remarks of **K. David Becker**, Chief Scientific Officer, Pathway Genomics Corporation, and **Elissa Levin**, Director, Genetic Counseling Program, Navigenics, Inc.

## HUMAN RESEARCH SUBJECTS, INSTITUTIONAL REVIEW BOARDS, AND INFORMED CONSENT

The DTC genetic testing company Navigenics is collaborating with the Scripps Translational Science Institute in an ambitious study that will follow 10,000 participants for 20 years, researching the predictive value of genetic risk markers, looking at outcomes, and evaluating the effectiveness of behavior modification. Although the Scripps IRB reviewed and approved the collaboration—including Navigenics' role—DTC genetic testing companies' research is not necessarily overseen by IRBs.[2]

Whether a genetic test is provided direct-to-consumer or in a more traditional clinical setting, ensuring that the client fully understands the test's risks as well as its benefits is critical. One potential risk, of course, is loss of genetic privacy and the many types of personal, emotional and financial harm that could ensue. A document designed to ensure that a client is giving free and informed consent must include an adequate description of the policies and procedures for securely handling the samples and the resulting data—including collecting, identifying and de-identifying, storing, using, sharing and destroying them. The factors and considerations that make up informed consent are many and complex. Is the customer, for example, merely consenting to have the DTC genetic testing done? Or is the individual also consenting to have the test results used for research? May a client consent to have the testing done without having his or her results used in research? One DTC company, in trying to streamline and clarify its informed consent document without weakening its protections, has so far managed to reduce the document to "only" six pages.[3]

A complicating factor relates to individuals who are unable to give informed consent—either test subjects who are minors, or a test subject's relatives—but whose lives could nevertheless be affected by the results. In China, for approximately $900, parents in Chongqing, PRC, can send their children—ages three to 12 years old—to a five-day camp for DNA testing to identify their gifts and talents, so they can focus on those strengths from an early age. According to the director of the Chongqing Children's Palace, "Nowadays, competition in the

---

[2] From the remarks of **Elissa Levin**, Director, Genetic Counseling Program, Navigenics, Inc.

[3] This sentence taken from the remarks of **Scott Woodward**, Director, Sorenson Molecular Genealogy Foundation.

"What does all this mean for the future of the perhaps archaic concept that we still teach in medical school called the therapeutic relationship? ... When does a client enter into a therapeutic relationship with you, with your company, with the employees of your company?"

Jonathan D. Moreno, Ph.D.
David and Lyn Silfen University Professor
Center for Bioethics, University of Pennsylvania Health System

"Is this actually the practice of medicine or not? If we are going to have these companies that actually are going to be putting some reports out providing risk assessment, even though you only do the risk calculation, you are providing information where clinical decisions will be made."

Andrea Ferreira-Gonzalez, M.D.
Professor of Pathology, Virginia Commonwealth University (VCU) and
Director, The Molecular Diagnostics Laboratory, VCU Health System

"Genetic information actually has a kinship relatedness to it... your close kin are also being dragged into this society without even knowing it, let alone consenting to it."

David Korn, M.D.
Vice Provost for Research
Harvard University

world is about who has the most talent. We can give Chinese children an effective, scientific plan at an early age."[4] In contrast, the framework of principles developed by the United Kingdom recommends that genetic testing of minors should only be carried out if there is a specific medical indication and if delaying the test till the age of consent might adversely affect that individual's medical care.[5] A similar view seems to be the norm in the United States. However, the issue of protecting "innocent bystanders"—customers' family members and extended kin-

---

[4] Chang E., In China, DNA tests on kids ID genetic gifts, careers, August 5, 2009. *CNN* (Available online at www.edition.cnn.com/2009/WORLD/asiapcf/08/03/china.dna.children.ability/, Accessed: March 29, 2010. Presented by Sandra Soo-Jin Lee, Senior Research Scholar, Stanford Center for Biomedical Ethics.

[5] From the remarks of **Timothy Aitman**, Professor of Clinical and Molecular Genetics, Division of Clinical Sciences, Imperial College London.

ship, who may not want their own genetic information revealed, even to themselves—will not be so easily addressed.

## QUESTIONS RAISED FOR FURTHER DISCUSSION

• When should DTC genetic tests be considered diagnostic, and when are DTC genetic testing companies practicing medicine?

• How can immediate or extended family members of tested individuals be protected from learning their genetic information if they choose not to, or from the effects of others' learning about it?

• Are DTC genetic testing companies subject to CLIA, HIPAA, and various state laws regulating clinical laboratories or the practice of medicine or not? This needs to be clarified, and consumers need to know and to understand the implications.

• Under what circumstances should DTC genetic testing customers be considered human research subjects and how should they be protected?

• Should DTC genetic testing companies use a common informed consent document (or a tailored version thereof), and if so, how should it be structured and worded so that it is comprehensive, understandable, clear and simple, and adequately protective?

• Should research conducted by DTC genetic testing companies, which is not covered by the Common Rule protecting human subjects and therefore not required to undergo review by an IRB, be required to undergo review by an IRB?

# Impact on Health Care
# and Public Health

DTC genetic testing may rapidly become a force in the health care system. The challenge is ensuring that it matures into a positive, constructive force, rather than a destructive, destabilizing one.

The medical value of genetic testing in certain clinical settings is indisputable. The ability to identify the individuals for whom more frequent screening is warranted—for breast or colon cancer, for example—may significantly reduce mortality from those disorders, enabling earlier diagnosis and treatment in some cases or aggressive prevention, such as prophylactic mastectomies or colectomies, in others. In many cases, these early interventions can significantly reduce not only human suffering, but might reduce health care costs as well.

What is not clear yet, however, is the societal economic value of DTC genetic testing and its corresponding impact on the medical system.

## ECONOMICS[1]

DTC genetic testing might lower overall health care costs by:

• Encouraging behavior changes and other prevention strategies: This is more likely for highly penetrant genes such as BRCA1 and BRCA2, whereas low-risk genes from a GWAS may have less impact;[2]

---

[1] Unless otherwise noted, the section "Economics" reflects the remarks of **Kathryn Phillips**, Professor of Health Economics and Health Services Research, University of California, San Francisco.

[2] This sentence taken from the remarks of **Patricia Ganz**, Professor of Health Services, School of Public Health and Professor of Medicine, David Geffen School of Medicine, University of California, Los Angeles.

- Increasing the prevalence of early detection and intervention;
- Shifting the access point for genetic testing from physicians to the Internet; or
- Facilitating pharmacogenomics, which might both lower costs and improve outcomes through better targeting of drugs.

Or, DTC genetic testing could raise overall health care costs if:

- More people decide to get tested;
- It results in more tests, follow-on screenings and interventions—many of which would almost certainly be unnecessary, given the current limitations of GWAS; or
- A negative test result gives a false sense of security and either leads to under-use of regular check-ups and other preventive measures, or is viewed as permission to resume, continue or commence smoking, poor diet, or a sedentary lifestyle—all behaviors that contribute to a multitude of chronic, multi-billion-dollar-a-year diseases.

In light of the potential for overuse of DTC genetic testing, it will be important to determine under what circumstances genetic testing is appropriate and beneficial. The World Health Organization's long-standing recommendations on disease screening[3]—which include that the condition should be an important health problem and there should be a treatment available for it—are a good place to begin. In addition, the test must be reliable, sensitive and specific; it must have strong positive and negative predictive values and be actionable.

## PHYSICIAN AWARENESS AND EDUCATION

DTC genetic testing companies generally advise clients to discuss test results with their health care providers—the majority of whom are not yet equipped to interpret the results and advise their patients. When tests have known value in clinical medicine (e.g. prenatal testing), physicians order them. And for known genetic conditions (cancer predisposition, for example), many physicians have at least a limited ability to interpret results. Most, however, lack the knowledge and expertise required to discuss the results of a GWAS. As a result, a

---

[3] Wilson MMG, Jungner G., Principles and practice of screening for disease. *WHO Chronicle* 1968; 22(11): 473.

patient may demand screening tests that have no proven value, or may consider the unsympathetic physician as behind the times and lacking in knowledge.[4]

A 2008 online survey of 1,880 physicians (including internists, pediatricians, obstetrician/gynecologists, and family-practice physicians) found fewer than half (42 percent) aware of DTC genetic testing. Among those who were aware, 42 percent had been asked by patients about DTC genetic testing, and 15 percent had patients who had brought in test results for discussion. Interestingly, in light of questions posed earlier about clinical utility, 75 percent of physicians who had discussed DTC genetic test results with their patients changed some aspect of the patient's care, such as screening tests offered, medications or dosages prescribed, lifestyle changes recommended, frequency of follow-up appointments, or diagnoses made.[5]

Clearly, institutions of medical education, including professional societies, need to engage fully in this matter to help physicians increase their knowledge and understanding of genetics so they can discuss the pros, cons and limitations of DTC genetic tests. The question is, "how?" The 2008 survey also asked physicians to rate the five sources they trusted most for patient health-related information. Journal articles scored highest, appearing in 96 percent of respondents' top five. Other high-scoring sources were government agencies (83 percent), other physicians (80 percent), professional organizations (74 percent) and medical web sites (62 percent); only 2 percent considered the media a trusted source. Asked where they were most likely to get information about DTC genetic testing, however, respondents overwhelmingly cited the media—reflecting a dearth of information available from more trusted sources.[6]

The Health & Human Services Secretary's Advisory Committee on Genetics, Health and Society (SACGHS) and numerous journal articles

---

[4] From the remarks of **Patricia Ganz**, Professor of Health Services, School of Public Health and Professor of Medicine, David Geffen School of Medicine, University of California, Los Angeles

[5] Kolor K, Liu T, St Pierre J, Khoury M., Health care provider and consumer awareness, perceptions, and use of direct-to-consumer personal genomic tests, United States, 2008, *Genetics in Medicine*. 2009, 11(8):595.

[6] This paragraph is based on the remarks of **Katrina Goddard**, Senior Investigator, Kaiser Permanente Center for Health Research.

have all lamented the low level of awareness and understanding among physicians.[7] For example,

- SACGHS found that "[Practitioners] cannot keep up with the pace of genetic tests [and] are not adequately prepared to use test information to treat patients appropriately . . . Practice guidelines are insufficient to ensure appropriate care."[8]
- Scheuner *et al.* wrote in the *Journal of the American Medical Association* that "the primary care workforce, which will be required to be on the frontlines of the integration of genomics into the regular practice of medicine, feels woefully unprepared to do so."[9]
- In an article in *Nature Reviews Genetics* [10] on challenges to genetics education for physicians and other health professionals, Guttmacher *et al* listed a crowded curriculum, misconceptions about genetics, lack of knowledgeable faculty, a disconnect between basic sciences and clinical experiences during training, inadequate representation of genetics on certifying exams, and failure to integrate genetics across the curriculum.
- In *Genetics in Medicine*, Suther and Goodson[11] listed a variety of challenges to integrating genetics into primary care: a dearth of genetics professionals, lack of knowledge about genetics among primary-care providers, lack of confidence, lack of referral guidelines, difficulty interpreting genetic tests, and difficulty explaining genetic risks to patients.

In general, medical schools have not embraced the challenge of adequately educating physicians in genetics and genomics. A 2007 publication of a survey [12] of accredited U.S. and Canadian medical schools found that although more than three-quarters teach medical genetics

---

[7] Unless otherwise noted, the remainder of the section "Physician Awareness and Education" is based on the remarks of **Joseph McInerney**, Executive Director, National Coalition for Health Professional Education in Genetics.

[8] *U.S. System of Oversight of Genetic Testing: A Response to the Charge of the Secretary of Health and Human Services*, Report of the Secretary's Advisory Committee on Genetics, Health, and Society, April 2008.

[9] Scheuner M, Sieverding P, Shekelle P., Delivery of genomic medicine for common chronic adult diseases: a systematic review**,** *JAMA*, 2008; 299(11):1320-1334

[10] Guttmacher A, Porteous M, McInerney J., Educating health care professionals about genetics and genomics, *Nature Reviews Genetics*, 2007;8(2):151-7

[11] Suther S, Goodson P., Barriers to the provision of genetic services by primary care physicians: a systematic review of the literature, *Genetics in Medicine*, 2003;5(2):70-6

[12] Thurston V, Wales P, Bell M, Torbeck L, Brokaw J., The current status of medical genetics instruction in U.S. and Canadian medical schools, *Academic Medicine*, 2007; 82 (5):441-445

to first-year students, fewer than half of the schools incorporate it into the 3rd and 4th years—the years that focus on clinical practice. Among medical schools that teach medical genetics, the overwhelming majority teach general concepts rather than practical applications, and only a little more than half integrate the topic into other courses.

Only two medical schools—in the University of Vermont and Johns Hopkins University—currently incorporate genetics training into all four years of medical school. Even if this became the norm, however, it will be decades before medical schools produce a critical mass of physicians with the knowledge and confidence to be as comfortable with medical genetics as they will be with the more traditional aspects of health care.

A nearer-term solution, then, needs to consider the following:

• *Whether geneticists will need to become familiar with all medical specialties or whether specialists will need to become conversant in genetics.*

   • In the United Kingdom, a parliamentary report recently concluded that it would be necessary for specialists to become conversant in genetics;[13] there has been no such deliberation or decision yet in the United States.

   • A related suggestion is to stop using the terms "genetic disorder" and "genetic disease," which perpetuates the mistaken impression that genetics is separate and distinct from the rest of medicine.[14]

• *How to provide practicing physicians with the information they need now, or will need tomorrow or next week.*

   • Suggestions being considered include continuing medical education courses, grand rounds, and making the information available online at the point of care, similar to the *Physician's Desk Reference*.[15]

   • Unfortunately, according to Levy and colleagues in *Genetics*

---

[13] From the remarks of **Timothy Aitman**, Professor of Clinical and Molecular Genetics, Division of Clinical Sciences, Imperial College London

[14] From the remarks of **Joseph McInerney**, Executive Director, National Coalition for Health Professional Education in Genetics

[15] www.PDR.net

*in Medicine,*[16] "many World Wide Web databases do not answer clinical questions about genetic conditions accurately. None of the resources we tested are efficient enough for point-of-care use. As genetics becomes more prominent in daily patient care, providers will need an efficient, accurate, and accessible source of information."

• In the United States, physician education will likely fall to the various professional societies. In contrast, the UK's National Genetics Education Development Center is responsible for educating National Health Service (NHS) providers about genetics. To date, however, the center has dealt primarily with single-gene disorders; it has recently begun to address the enormous need to educate the NHS work force on the genetics of common diseases.[17]

In the long run, many participants agreed that it appears that the health care system will need more primary care physicians knowledgeable about genomics and more genetic counselors.

## GENETIC COUNSELING[18]

Genetic counseling is only beginning to emerge from the common misperception that it is primarily about inheritance of single-gene disorders and therefore is largely the province of obstetrician/gynecologists or specially trained genetic counselors and pediatricians.

Genetic counseling is: "the process of helping people understand and adapt to the medical, psychological, and familial implications of genetic contributions to disease." Counselors interpret family and medical histories; educate clients about inheritance, testing, management, prevention, resources and research; and counsel them to promote informed choices and adaptation to the risk or condition.[19] Counseling is often at least as

---

[16] Levy H, LoPresti L, Seibert D., Twenty questions in genetic medicine—an assessment of World Wide Web databases for genetics information at the point of care. *Genetics in Medicine.* 2008;10(9):659-67.

[17] From the remarks of **Timothy Aitman**, Professor of Clinical and Molecular Genetics, Division of Clinical Sciences, Imperial College London.

[18] Unless otherwise noted, the section "Genetic Counseling" reflects the remarks of **Joseph McInerney**, Executive Director, National Coalition for Health Professional Education in Genetics.

[19] Resta R, Biesecker B, Bennett R, Blum S, Hahn S, Strecker M, Williams J., A new definition of genetic counseling: National Society of Genetic Counselors' Task Force report. *Journal of Genetic Counseling.* 2006;15(2):77-83.

> "... we are dealing with a disruptive technology that may truly revolutionize the way we practice medicine. But we have to integrate this technology using evidence based practice to show its added value to our current health care system."
>
> Muin J. Khoury, M.D., Ph.D.
> Director, Office of Public Health Genomics,
> Centers for Disease Control and Prevention and
> Senior Consultant in Public Health Genomics,
> National Cancer Institute

vital before a genetic test as it is afterwards; it puts the test in perspective and helps prepare clients for potential findings.

Some DTC genetic testing companies offer genetic counseling—via telephone or the Internet, or sometimes in person—as part of their testing services. Some have counselors on staff, and others contract with specific counselors and refer clients to them. And at least one charges its customers for the service—$250 per hour.[20] Several DTC testing companies currently offer genetic counseling services to their customers at no additional charge, so there is no cost barrier. It appears, however, that clients often are not aware that these services are available—or even what genetic counseling is—and thus do not always benefit from them.[21]

Billing and reimbursement are already problematic for genetic counseling in connection with testing in a traditional clinical setting. As DTC testing becomes more widespread and the industry evolves, reimbursement may become an issue in this context as well.[22]

## QUESTIONS RAISED FOR FURTHER DISCUSSION

• How can health care systems, in the United States and elsewhere, ensure that the enormous potential of genetic testing generally—and DTC testing in particular—is harnessed and directed to maximize its benefits for human health?

• What are the most effective ways to incorporate genetics into

---

[20] From the remarks of **Patricia Ganz**, Professor of Health Services, School of Public Health and Professor of Medicine, David Geffen School of Medicine, University of California, Los Angeles.

[21] From the remarks of **Elissa Levin**, Director, Genetic Counseling Program, Navigenics, Inc.

[22] Concern voiced by a workshop attendee during a question and answer period.

medical education and training—in medical school, in postgraduate (residency and fellowship) training, in continuing medical education, and/or at the point of care?

• In light of the general public's lack of information in the areas of health literacy and genetics and in particular of the role, risks, and benefits of genetic testing in health care, what are the most effective ways to educate the public in these areas? Would multiple education strategies best serve the public interest?

• How does framing affect the public's perception of the risks/benefits of genetic testing?

• What are the roles and appropriate levels of genetic counseling under various circumstances of genetic testing?

• How will DTC genetic testing, pre-test and post-test counseling, and follow-up care be paid for?

# Current Legislative and Regulatory Framework in the United States[1]

## FEDERAL LEGISLATION RELEVANT TO DIRECT-TO-CONSUMER GENETIC TESTING

The following federal laws—the 1976 Medical Device Amendments to the Food, Drug, and Cosmetic Act, the Clinical Laboratory Improvement Amendments, the Health Insurance Portability and Accountability Act, and the Genetic Information Nondiscrimination Act—are most relevant to the oversight of DTC genetic testing. At the present time, however, only the latter explicitly applies to such tests.

**Food, Drug, and Cosmetic Act (FDCA)** and **Medical Device Amendments (MDA)**. Enacted in 1938, FDCA brought the sale of foods, drugs, and cosmetics under the purview of the Food and Drug Administration (FDA). The Act prohibited false therapeutic claims and required manufacturers to demonstrate the safety of their products under stipulated conditions of use before they could be marketed.[2] FDCA placed the regulation of medical devices within the purview of the FDA. It was only with the 1976 Medical Device Amendments to FDCA, however,

---

[1] The section "Current Legislative and Regulatory Framework in the United States" is based on the remarks of **Andrea Ferreira-Gonzalez**, Professor of Pathology, Virginia Commonwealth University and Director, The Molecular Diagnostics Laboratory, Virginia Commonwealth University Health System; and **Courtney Harper**, Acting Director of the Division of Chemistry and Toxicology Devices, Office of *In Vitro* Diagnostic Device Evaluation and Safety, Center for Devices and Radiological Health, U.S. Food and Drug Administration.

[2] Department of Health and Human Services (HHS), FDA's Origins & Functions, Available at: http://www.fda.gov/AboutFDA/WhatWeDo/History/Origin/default.htm, Accessed: July 16, 2010.

that the term "medical device" was explicitly defined. Under this definition, diagnostic tests are classified as medical devices.

**Clinical Laboratory Improvement Amendments (CLIA).** Enacted in 1988, CLIA establishes quality standards for all non-research laboratory testing performed on human specimens for the purpose of providing information for diagnosing, preventing or treating disease or for assessing health.[3] Laboratories performing these tests must be certified by the Centers for Medicare and Medicaid Services. Although CLIA is a federal law, two states—New York and Washington—are exempt from CLIA because their state-level regulations have been determined to meet or exceed CLIA's requirements.

**Health Insurance Portability and Accountability Act (HIPAA).** Enacted in 1996 to protect health insurance coverage for workers—and their families—who change or lose their jobs, a major provision of HIPAA was to require the Department of Health and Human Services to write regulations aimed at protecting the privacy of individuals' identifiable personal health information should Congress fail to enact legislation offering such protection before August 31, 1999.[4]

**Genetic Information Nondiscrimination Act (GINA).** Enacted in 2008 and in force since 2009, GINA[5] protects the public from discrimination based on genetic information in health insurance and employment settings. It prohibits employers from requesting, requiring, purchasing or using genetic information about an individual or family member in any job-related decision. This codifies the Equal Employment Opportunity Commission's interpretation of the Americans with Disabilities Act—never tested in court—by prohibiting workplace discrimination based on genetic information. It also prohibits health insurers (both group and individual) from requesting, requiring or using a person's

---

[3] The Centers for Medicare and Medicaid Services, CLIA Overview. Available at: www.cms.hhs. gov//clia/, Accessed: March 29, 2010.

[4] U.S. Department of Health and Human Services, Health Information Privacy: The Health Insurance Portability and Accountability Act of 1996 (HIPAA), Available at: www.hhs.gov/ocr/ privacy/, Accessed: March 29, 2010.

[5] Department of Health and Human Services (HHS), "GINA," The Genetic Information Nondiscrimination Act of 2008: Information for Researchers and Health Care Professionals, April 6, 2009. Available at: www.genome.gov/Pages/PolicyEthics/GeneticDiscrimination/GINAInfoDoc. pdf, Accessed: March 29, 2010.

genetic information for underwriting purposes. In contrast, it does not prohibit insurers' use of such information in underwriting life, disability or long-term care insurance.

Although some DTC genetic testing companies maintain that their laboratories are CLIA-certified and that they abide by HIPAA's privacy provisions, it is not clear that the companies are legally subject to CLIA. Futhermore, they do not fall within the definition of "covered entities" subject to HIPAA. And, though many clients of DTC genetic testing companies may not even be aware of these laws, those who are familiar with them would likely assume that their protections apply to DTC genetic testing.

## REGULATING DIRECT-TO-CONSUMER GENETIC TESTING AT THE FEDERAL LEVEL

The FDA is the federal agency charged with regulating genetic tests—whether they are offered direct-to-consumer or provided in a clinical setting.

Although the FDA has primary oversight of genetic testing, other agencies occasionally become involved. The Federal Trade Commission (FTC) has jurisdiction over claims made in the advertising and promotion of any commercial product, genetic tests included. In 2006, the Senate Special Committee on Aging asked the Government Accountability Office (GAO) to investigate and report on so-called "nutrigenetic" testing web sites. Undercover agents from the GAO's Forensic Audits and Special Investigations unit found not only that the four companies examined made medically unproven and meaningless health-related predictions, but also that they apparently had not even analyzed the DNA samples that the investigators had submitted for testing.[6]

The 2008 SACGHS report[7] identified more than two dozen gaps in oversight within and among the various components of the current

---

[6] U.S. Government Accountability Office, Testimony Before the Special Committee on Aging, U.S. Senate, Nutrigenetic Testing: Tests Purchased from Four Web Sites Mislead Consumers, Statement of Gregory Kutz, Managing Director Forensic Audits and Special Investigations, Report available at: http://www.gao.gov/new.items/d06977t.pdf, Accessed: March 29, 2010.

[7] *U.S. System of Oversight of Genetic Testing: A Response to the Charge of the Secretary of Health and Human Services*, Report of the Secretary's Advisory Committee on Genetics, Health, and Society, April 2008.

system. Two gaps of particular concern relate to proficiency testing (PT) and laboratory-developed tests (LDTs).

*Proficiency testing.* Laboratories seeking to earn or renew CLIA certification are judged on how well they perform specific analytical tests. Current regulations require formal proficiency testing for only 82 analytes, none of which is relevant in genetic testing. CLIA permits alternative assessments in such cases, but these are not always feasible or acceptable—particularly in as new and rapidly evolving a field as genetic testing—and to date the benefit of alternative assessments has not been systematically demonstrated. However, all clinical tests require some form of assessment whether they are specifically identified in CLIA or not.[8]

*Laboratory-developed tests (LDTs).* Under MDA and successor amendments, medical devices are grouped into three classes (Class I, Class II, and Class III) with increasing regulatory requirements aimed at assuring that the various types of devices are safe and effective before they are marketed[9]—Class I devices are classified as low risk, Class II devices as moderate risk, and Class III devices as high risk. The FDA generally considers laboratory-developed tests to be Class II devices. Devices in this class require FDA clearance largely on the basis of demonstrating "substantial equivalence" to a legally marketed "predicate device." The FDA may require clinical data from the manufacturer to establish both analytical and clinical performance. In addition, the FDA may impose special controls on the device when it goes to market. These include special labeling requirements, mandatory performance standards, and post market surveillance. Under a technicality known as "enforcement discretion," the FDA has chosen not to require such testing and approval for an LDT developed in a CLIA-certified laboratory and used only by that particular laboratory. Many novel genetic tests, including DTC tests, fall into this category.[10]

Although some DTC genetic testing companies describe their laboratories as meeting all of CLIA's standards, they currently are not required

---

[8] [42 CFR 493.1253] Standard: Establishment and verification of performance specifications.

[9] Department of Health and Human Services, Medical Devices, Available at: http://www.fda.gov/MedicalDevices/default.htm, Accessed: September 10, 2010.

[10] There is significant disagreement, however, about what constitutes the appropriate regulatory status of LDTs.

"I do think [DTC genetic testing companies] are engaging in a practice that is getting very, very close to [that of] a clinical laboratory. I think that there ought to be a uniform system of standards that are applied both to the commercial sector and to the health care sector, so that the people who receive the information have the same level of confidence—or whatever word you want to use—in the results."

David Korn, M.D.
Vice Provost for Research
Harvard University

to do so. Moreover, a bill pending in the California State Senate would specifically exempt DTC genetic testing companies from CLIA.[11]

The FDA is currently developing a report on oversight of DTC genetic testing and has recently acted to regulate DTC tests (see footnote 3). **Appendix B** describes how DTC genetic testing is regulated in the United Kingdom and elsewhere outside the United States.

## QUESTIONS RAISED FOR FURTHER DISCUSSION

- How should current federal laws and regulations, particularly the MDA, CLIA, and HIPAA, apply to DTC genetic testing?
- Are additional laws and regulations needed?
- How should the regulatory gaps—including the concerns raised by the lack of required proficiency testing and the lack of regulatory oversight of laboratory-developed tests—be closed?

---

[11] California SB 482 Senate Bill, Available at: http://info.sen.ca.gov/pub/09-10/bill/sen/sb_0451-0500/sb_482_bill_20090414_amended_sen_v98.html, Accessed: May 24, 2010.

# Areas for Further Study

As stated at the outset, it would be impossible in a one and a half day workshop to identify and explore all the issues raised by DTC genetic testing—much less to address them adequately. The workshop participants did, however, identify a number of areas where further study appears to be warranted. These areas include the impact on individuals, public health, the health care system and medical research, along with the implications for legislation and regulation.

## IMPACT ON INDIVIDUALS

• Is the genetic privacy of individuals sufficiently protected by current laws and regulations, both generally and in light of the burgeoning DTC genetic testing industry?

• Should the use of genetic information in underwriting life, disability and long-term care insurance be prohibited or otherwise regulated—and if so, how can this be done while balancing the legitimate business needs of insurers?

## IMPACT ON PUBLIC HEALTH

• How can DTC genetic testing be channeled and used to protect and improve public health?

• Can and should DTC test results that indicate even a slightly increased risk (of lung cancer, for example) be used to change unhealthy behaviors (encourage smoking cessation, for example)?

• Conversely, how can this be done without potentially enabling unhealthy behaviors among individuals whose test results indicate no increase—or even a slight decrease—in risk?

## IMPACT ON THE HEALTH CARE SYSTEM

• How can DTC genetic testing improve health care quality, access and outcomes without significantly adding to health care costs? Can it help lower costs by stimulating behavioral changes?

• How can overuse and inappropriate use of DTC testing be prevented?

• What will be the impact of supply and demand on the number of primary care providers and genetic counselors?

• How should the medical education system change to enhance the genetics capability and proficiency of all physicians?

• What are the most effective ways to provide ongoing medical education and point-of-care information?

Is this just something people have a right to choose, just like they could choose in any market to purchase any product? . . . We have a situation where public understanding, much less clinical understanding, is wanting. . . . Education, strategies for research, opportunities for learning about the uses and misuses, the role of promotion in terms of transparency and accountability, the relationship of private initiative to public responsibility . . . the relevant determinants of appropriate regulatory response all framed in a dynamic environment where we can expect more will be learned increasingly over time. . . These decisions are potentially life-influencing if not life-threatening and [have] a very high consequence potentially for individual economic as well as physical well-being. . . When you couple those with the profoundly individual personal character of this information . . . the meaning of this for each individual and the potential risk of abuse, I think this is qualitatively distinct and deserving of special attention. . . This is driving toward a larger convergence of quality improvement in care, clinical decision support, continuing medical education and research, all coming together. . . driven by the same underlying database . . . applied to the needs of that individual patient. . . They are likely to evolve in an accelerating way where those elements of interaction or vulnerability that have been validated become an expected part of the safety profile and high-quality care delivery. To me, that's not only thinkable, I think that's almost inevitable.

Harvey Fineberg, M.D., Ph.D.
President, Institute of Medicine

## IMPACT ON MEDICAL RESEARCH

• How can the enormous data resources of DTC genetic testing companies be most usefully directed to benefit patients and support genetics research overall?

• What is to prevent genetic data collected for genealogical purposes from being used in health risk assessments?

• At what point do DTC genetic testing companies' customers become human research subjects? Should DTC genetic testing companies be subject to the protections for human research subjects specified in federal regulations?

• What will an appropriate informed consent document look like?

## IMPLICATIONS FOR LEGISLATION AND REGULATION

• Does the FDA have sufficient authority to regulate DTC testing or are new laws or regulations needed?

• How can the DTC industry be regulated sufficiently to protect consumers, without unduly stifling innovation and investment?

# Appendixes

# Appendix A

# Glossary Of Terms And Acronyms

**allele.** Any one version of a particular gene, occupying the same position on a specific chromosome as other alleles of that gene and differing from the others by only one or a few nucleotides. Each human cell contains two copies of each gene; these may be identical, or they may be two different alleles.

**BRCA1/BRCA2.** Breast cancer susceptibility genes. These two genes normally produce proteins involved in DNA repair. Mutations in these genes are associated in certain populations with a high risk for breast cancer and a significantly increased risk for ovarian cancer—with or without a family history of those disorders. Specific **alleles** of BRCA1 and BRCA2 genes have been associated with increased risk of developing breast (60% vs. 12% general population rate) and ovarian (15-40% vs. 1.4% general population rate) cancers. Together, BRCA1 and BRCA2 mutations account for between 5 and 10 percent of all breast cancers.

**CDC.** Centers for Disease Control and Prevention, Department of Health & Human Services.

**CLIA.** Clinical Laboratory Improvement Amendments of 1988.

**CMS.** Centers for Medicare and Medicaid Services, Department of Health & Human Services.

**complex disorder.** A disorder associated with a combination of **alleles** of several different genes—in contrast with **Mendelian**, single-gene, disorders. Because of the prevalence of many complex disorders in the general population, they are sometimes referred to as "common diseases."

**dominant.** In an individual with two different **alleles** for a particular gene, the dominant allele determines the observed **phenotype**.

**epigenetics.** Changes in an organism's **phenotype** by changes in gene expression that are not a result of the organism's genome sequence.

**FDCA.** Food, Drug, and Cosmetic Act of 1938.

**GINA.** Genetic Information Nondiscrimination Act of 2008.

**GWAS.** Genome-wide association study.

**genotype.** An individual organism's entire, exact genetic makeup, including all its **alleles**—regardless of whether or not those alleles are expressed.

**HIPAA.** Health Insurance Portability and Accountability Act of 1996.

**haplotype.** A set of **alleles** on a chromosome that tend to be inherited together.

**HapMap.** A **hap**lotype **map** of a consensus human genome. The HapMap is a catalog of common genetic variants that occur in human beings. It describes what these variants are, where they occur in human DNA, and how they are distributed among people within populations and among populations in different parts of the world.

**heterozygous.** Having two different variations (**alleles**) of a DNA sequence.

**highly penetrant.** See **penetrance**.

**homozygous.** Having two identical **alleles** for a particular DNA sequence.

**informed consent.** An ethical and legal requirement in both clinical practice and in research with human participants that is necessary to ensure that an individual patient or client is aware of the risks and benefits of participating in research or of undergoing a clinical procedure. The individual must be informed of the nature of the procedure, possible alternatives, and any costs and potential risks as well as benefits.

**LDT.** Laboratory-developed test.

**MDA.** Medical Device Amendments of 1976.

**Mendelian disorder**. A disorder that has a strong association with a single gene. Also called a single-gene disorder.

**NHGRI.** National Human Genome Research Institute, an institute of the National Institutes of Health.

**NICE**: National Institute for Health and Clinical Excellence (United Kingdom).

**nutrigenomics:** The study of how dietary constituents interact with specific genes to affect health and risk of disease.

**penetrance.** Within the universe of individuals who carry a particular **allele**, the proportion that actually express a phenotype associated with that allele.

**phenotype.** The observable physical characteristics of an organism, as determined by its genetic makeup, epigenetic modifications, and typically, by environmental influences.

**recessive.** The traits encoded by a recessive **allele** in a particular gene will only be physically apparent if it is present in both copies of the gene and cannot therefore be "over-ridden" by a dominant allele.

**SACGHS.** Secretary's Advisory Committee on Genetics, Health, & Society. (Refers to the Secretary of Health & Human Services.)

**SNP.** Single-nucleotide polymorphism. A variation within a genome wherein a single nucleotide differs from the consensus genome.

# Appendix B

# Regulation of Direct-to-Consumer Genetic Testing Outside the United States

## RELEVANT LEGISLATION AND REGULATORY AND ADVISORY BODIES IN THE UNITED KINGDOM[1]

The United Kingdom (UK) does not yet have a legal or regulatory structure in place for complex gene tests or the DTC genetic testing companies that offer them. However, a collection of UK laws, regulatory authorities and advisory bodies, as well as European Union (EU) directives, are concerned with genetic testing in general.

### UK Legislation

Genetic testing in the UK—and throughout the EU—is subject to the EU's 1998 *In Vitro* Diagnostic Devices Directive (IVDD).[2] In the UK specifically, the 2003 report *Genes Direct*[3] was the impetus for the UK's Human Tissue Act of 2004, which makes DNA theft—the collection or analysis of individuals' DNA without their consent—a crime, punishable with a fine or imprisonment up to three years.

### UK Regulatory Authority and Advisory Bodies

The UK's chief medical regulatory agency is the Medicines and Healthcare Products Regulatory Agency. Among other responsibili-

---

[1] Unless otherwise noted, the section "Relevant Legislation and Regulatory and Advisory Bodies in the United Kingdom" is based on the remarks of **Timothy Aitman**, Professor of Clinical and Molecular Genetics, Division of Clinical Sciences, Imperial College, London.

[2] Report available from: www.mhra.gov.uk/Howweregulate/Devices/InVitroDiagnosticMedicalDevicesDirective/index.htm, Accessed: March 29, 2010.

[3] Report available at: www.hgc.gov.uk/UploadDocs/DocPub/Document/genesdirect_full.pdf, Accessed: March 29, 2010.

ties, the Agency is charged with implementing the European Union's IVDD, which encompasses genetic testing. Advisory bodies include the Genetic Testing Network and the Human Genetics Commission, which is the UK government's main advisory body on genetics. In addition, the National Institute for Health and Clinical Excellence advises on the clinical utility and effectiveness of new medical products, including genetic tests.

The Genetic Testing Network evaluates tests for Mendelian diseases, according to the well-established gene dossier protocol. There currently is no analogous formal body for common diseases.

The Human Genetics Commission (HGC) began considering DTC genetic tests in the 2003 report *Genes Direct*. Among other recommendations, the report advised that genetic testing for Mendelian diseases should not be offered direct-to-consumer, but only in the traditional clinical setting and with adequate pre- and post-test counseling. The HGC produced an updated report *More Genes Direct* in 2007.[4] The report determined that most genetic tests fall into the lowest of four risk categories and therefore do not require pre-market assessment of clinical validity or utility; self-certification is sufficient. The Commission has recently revisited its assessment and has now recommended that genetic testing should be classified as medium-risk. The European Union's IVDD still considers genetic testing to be low-risk, however, and the UK's official position reflects the EU's assessment.

The HGC held a seminar in 2008 with a wide range of stakeholders, including some DTC genetic testing companies. One of the conclusions, strongly supported by the companies, was to develop a common international code of practice. In September 2009, the HGC released for comment a draft document, *A Common Framework of Principles for Direct to Consumer Genetic Testing Services*.[5] The framework, which could form the basis of an international code of practice, included the principles:

---

[4] Report available at: www.hgc.gov.uk/UploadDocs/DocPub/Document/More%20Genes%20 Direct.pdf, Accessed: March 29, 2010.

[5] Report available at: www.hgc.gov.uk/client/Content.asp?ContentId=816, Accessed: March 29, 2010. The HGC subsequently released its final report on August 4, 2010. The report, A Common Framework of Principles for direct-to-consumer genetic testing services, is available at: www.hgc. gov.uk/Client/document.asp?DocId=280&CAtegoryId=10, Accessed August 18, 2010.

• Test providers need to provide a high-quality service that both meets client expectations and safeguards their interests;

• Tests for inherited diseases should be provided only with individualized pre- and post-test counseling;

• Providers should comply with local, national or international advertising guidelines, and promotional claims should describe tests accurately and without bias;

• Providers should supply clients with easily understood information, including likely outcomes of the test;

• Tests should be carried out only with clients' free and informed consent; and

• Unless there is a clinically important reason to do so, individuals should not be tested until they have reached the age of consent.

Comments on the common framework were due back to the Commission in December 2009.

The UK Parliament has also begun considering issues raised by genetics and genomics. In July 2009, the Science and Technology Committee of the House of Lords published its *Genomic Medicine* report. The report recommended establishing:

• A voluntary code of practice, which encourages providers to be open about tests' limitations and enables consumers to make informed decisions; and

• A Department of Health web site for consumers containing up-to-date and comprehensive information about the DTC genetic testing companies and the tests they offer.

Because DTC genetic testing is a rapidly expanding international market and therefore not easily regulated using strictly national guidelines, the UK's Human Genetics Commission is eager to liaise with regulators in other countries.

## Beyond the United Kingdom

In April 2009, the German parliament passed legislation banning DTC genetic testing. Under this law, genetic tests can be administered in Germany only by licensed physicians and with patients' informed consent.

A comprehensive listing and description of other legislative and legal documents related to genetic testing in the European Union, EU-member and other European countries, and non-European countries can be found in the report, *Definitions of Genetic Testing in European and other Legal Documents*,[6] prepared for the European Commission in 2009.

---

[6] Report available at: www.eurogentest.org/web/files/public/unit6/core_competences/BackgroundDocDefinitionsLegislationV10-FinalDraft.pdf, Accessed: March 29, 2010.

# Appendix C

# Currently Available Direct-to-Consumer Genetic Tests

## DISEASE SUSCEPTIBILITY

Abdominal Aortic Aneurysm
Age-related Macular
    Degeneration
Alcohol Dependence
Alopecia Areata
Alzheimer's Disease (late-onset)
Ankylosing Spondylitis
Asthma
Atopic Dermatitis
Atrial Fibrillation
Attention-Deficit Hyperactivity
    Disorder
Back Pain
Basal Cell Carcinoma
Bipolar Disorder
Bladder Cancer
Brain Aneurysm
Brain Cancer (Glioma)
Breast Cancer
Celiac Disease
Chronic Obstructive Pulmonary
    Disease (COPD)
Cleft Lip and Cleft Palate
Cluster Headaches
Colorectal Cancer

Creutzfeldt-Jakob Disease
Crohn's Disease
Developmental Dyslexia
Diabetes (Gestational)
Diabetes (Type 1)
Diabetes (Type 2)
Endometriosis
Esophageal Cancer
Essential Tremor
Exfoliation Glaucoma
Follicular Lymphoma
Gallstones
Gout
Hashimoto's Thyroiditis
Heart Attack
Heart Disease
High Blood Pressure
    (Hypertension)
Infertility (Male)
Intervertebral Disc Disease
Intrahepatic Cholestasis of
    Pregnancy
Kidney Disease
Kidney Stones
Larynx Cancer

Lou Gehrig's Disease (ALS)
Leukemia (Chronic Lymphocytic Leukemia)
Lung Cancer
Lupus (Systemic Lupus Erythematosus)
Multiple Sclerosis
Myasthenia Gravis
Neural Tube Defects
Neuroblastoma
Nonalcoholic Fatty Liver Disease
Obesity
Obsessive-Compulsive Disorder
Oral and Throat Cancer
Osteoarthritis
Osteoporosis
Otosclerosis
Ovarian Cancer
Paget's Disease of Bone
Pancreatic Cancer
Parkinson's Disease
Periodontal Disease
Peripheral Arterial Disease (PAD)
Placental Abruption

Preeclampsia
Primary Biliary Cirrhosis
Progressive Supranuclear Palsy
Prostate Cancer
Psoriasis
Restless Leg Syndrome
Rheumatoid Arthritis
Rheumatoid Arthritis (Juvenile)
Sarcoidosis
Schizophrenia
Sjögren's Syndrome
Skin Cancer (Basal Cell Carcinoma)
Skin Cancer (Melanoma)
Stomach Cancer
Stroke
Tardive Dyskinesia
Testicular Cancer
Thrombosis
Thyroid Cancer
Tourette's Syndrome
Ulcerative Colitis
Uterine Fibroids

## TRAITS

ABO Blood Types
Alcohol Dependence
Alcohol Flush Reaction
Asparagus Metabolite Detection
Athletic Performance
Avoidance of Errors
Baldness (Common form)
Birth Weight
Bitter Taste Perception
Blood Glucose
Breastfeeding and IQ
C-reactive Protein Level
Cocaine Dependence

Compatibility (DNA Dating)
Earwax Type
Eye Color
Food Preference
Freckling
HDL Cholesterol Level
HIV Progression
Hair Color
Hair Curl
Hair Thickness
Height
Lactose Intolerance / Persistence
Leprosy Susceptibility

Longevity
Malaria Complications
Malaria Resistance (Duffy
    Antigen)
Measures of Intelligence
Measures of Obesity
Memory
Metabolic Health
Muscle Performance
Nicotine Dependence
Non-ABO Blood Groups
Norovirus Resistance

Nutritional Needs
Odor Detection
Pain Sensitivity
Persistent Fetal Hemoglobin
Photic Sneeze Reflex
Red Hair Color
Resistance to HIV/AIDS
Response to Diet and Exercise
Sex Hormone Regulation
Short Stature
Tuberculosis Susceptibility
Weight Loss

## CARRIER/DIAGNOSTIC TESTING

Abetalipoproteinemia (Selected
    Ashkenazi Jewish Specific
    Mutations)
Adrenal Hypoplasia, classic
    and non-classic (Selected
    Ashkenazi Jewish Specific
    Mutations)
Alpha-1 Antitrypsin Deficiency
Beta/Sickle Cell Disease
BRCA Cancer Mutations
    (Selected)
Bloom's Syndrome
Canavan Disease
Chorea-Acanthocytosis (Selected
    Ashkenazi Jewish Specific
    Mutations)
Connexin 26-Related Sensorineu-
    ral Hearing Loss
Cystic Fibrosis
Factor V Leiden
Factor XI Deficiency
Familial Dysautonomia
Familial Hypercholesterolemia

FamImal Hyperinsulinemia
    (Selected Ashkenazi Jewish
    Specific Mutations)
Familial Mediterranean Fever
Fanconi Anemia (Selected
    Ashkenazi Jewish Specific
    Mutations)
G6PD Deficiency
Gaucher Disease
Glycogen Storage Disease Type
    1a
Hemochromatosis
Limb-girdle Muscular Dystrophy
Maple Syrup Urine Disease Type
    1B
Mucolipidosis IV
Nemaline Myopathy (Selected
    Ashkenazi Jewish Specific
    Mutations)
Niemann-Pick Disease Type A
Phenylketonuria
Prothrombin II
Rhizomelic Chondrodysplasia
    Punctata Type 1 (RCDP1)

Sickle Cell Anemia & Malaria Resistance
Spinal Muscular Atrophy
Tay-Sachs Disease

Torsion Dystonia
Usher Syndrome, Types 1F and 3 (Selected Ashkenazi Jewish Specific Mutations)

## DRUG RESPONSE

Abacavir Hypersensitivity
Alcohol Metabolism
Antidepressant Response
Asthma Medication Response
Beta-Blocker Response
Caffeine Metabolism
Carbamazepine Response
Clopidogrel (Plavix®) Efficacy
Floxacillin Toxicity
Fluorouracil Toxicity
Heroin Addiction
Irinotecan Response
Naltrexone Treatment Response
Postoperative Nausea and Vomiting (PONV)

Pseudocholinesterase Deficiency
Response to Hepatitis C Treatment
Response to Interferon Beta Therapy
Simvastatin Response
Statin Response (Stain Induced Myopathy)
Succinylcholine Response
Thiopurines Response
Tamoxifen
Warfarin (Coumadin®) Sensitivity

## OTHER SERVICES

Ancestry Testing
Biological Relationship/Kinship testing

Genetic Counseling

*Information courtesy of AccessDNA® www.AccessDNA.com, August 2010.*

# Appendix D

# Representative Direct-to-Consumer Genetic Testing Companies

## PERSONAL GENOMICS

23andMe
Alio Genetics
Consumer Genetics
CygeneDirect
deCODEme
DNA Traits
GeneEssence
GeneSNP
Holistic Health International

Inherent Health
Matrix Genomics
MyGenome
MyRedHairGene.com
Navigenics
Respiragene
Salugen
Scientific Match
vuGene

## ANCESTRY OR BIOLOGICAL RELATIONSHIP TESTING

23andMe
Affiliated Genetics
African Ancestry
AllTests International, Inc.
Ancestry.com
ARCpoint
ARGUS BioSciences
Biosynthesis, Inc.
Boston Paternity
Cambridge DNA
Chromosomal Laboratories, Inc.
Consumer Genetics
DNA-CARDIOcheck
DNA Consultants

DNA Diagnostics Center
DNA Diagnostics, Inc.
DNA Dimensions
DNA Direct
DNA Heritage
DNA Lab Center
DNA Plus
DNA Profiles of America
DNA Reference Laboratory
DNA Roots
DNA Solutions
DNA Testing Centre
DNA Tribes
Determigene

easy DNA
EthnoAncestry
Family Genetics
Family Tree DNA
GFI Laboratory
GeneMatch
GeneSys Biotech
GeneTrack
Gene Tree
Genebase Systems
Genelex
Genomic Express
HealthServices

HomeDNA
IDENTIGene
Metaphase Paternity
Paternity Experts
PaternityTesters.com
Paternity Testing Corporation
Prenatal Genetics Center
Proactive Genetics
Prophase Genetics
Roots for Real
Test Country
The Genographic Project
Triad DNA

*Information courtesy of AccessDNA® www.AccessDNA.com, August 2010.*

# Appendix E

# Planning Group Biographies

## CO-CHAIRS

**Frederick R. Anderson, Jr.,** B.A., (History of Science), University of North Carolina; BA (Jurisprudence), Oxford University; J.D., Harvard Law School, is a partner of the law firm of McKenna, Long & Aldridge LLP in Washington, D.C. He addresses "green" technologies (wind, solar, fuel cell, biofuels), legal strategies for bringing chemicals to market, a wide variety of Clean Air Act issues, risk assessment and management, greenhouse gas science and international negotiations, chemical regulation, FDA and EPA food contact and antimicrobial approvals, access to genetic information, and environmental aspects of international trade. Mr. Anderson is former Dean of the law school at American University, and was the first full-time President of the Environmental Law Institute. He was the first Editor-in-Chief of the *Environmental Law Reporter* and was Chairman of the American Bar Association's Standing Committee on Environmental Law. He served on a 12-member Congressional study commission created by the Superfund legislation to examine toxic tort recovery for injury from hazardous substances. Mr. Anderson was a member of the Harvard Group on Risk Management and Reform and was chairman of the Advisory Working Group on Environmental Sanctions for the U.S. Sentencing Commission. He has been both a member of and a consultant to the Administrative Conference of the United States. In 1984 his analysis of the Superfund program resulted in the adoption by the Conference of a policy favoring negotiated solutions to disputes about waste site cleanup. He is also Chairman of the Board of the Institute for Governance and Sustainable Development and was chairman of the American Bar Association's Commission on Inter-American Affairs. As chairman of the ABA's Standing Committee on Environmental Law, he played a key role in organizing international conferences in Europe and Canada on acid rain, and in Mexico City on environmental issues. He was the founding chair of the board of the Center for International Environmental Law, a position he held for 19 years. His book, *NEPA in the Courts*, went through several printings and was a Literary Guild selection. His book, *Environmental Improvement through Economic Incentives*, has been translated and widely used by governments and in universities. He has written numerous scholarly articles as well as other pieces. Mr. Anderson has served on several National Academies committees, including (1) Board on Atmospheric Sciences and Climate, (2) Commission on Life Sciences, (3) Board on Environmental Studies and Toxicology, (4) Committee on Industrial Competitiveness and Environmental Protection, (5) Committee to Review Risk Management in the DOE's Environmental Remediation Program, (6) Panel on Integration of Socio-Economic Criteria into the Site Selection Process for a High-Level Radioactive Waste Repository, and (7) Committee on EPA Assessment Factors for Data Quality. He was a member of the National Academies' planning committee that initiated the 1997 Academy Symposium on Science, Technology, and Law. He has been a member of the Committee on Science, Technology, and Law since its inception.

**Barbara E. Bierer**, M.D., is Senior Vice President for Research at the Brigham and Women's Hospital and Professor of Medicine at Harvard Medical School. Dr. Bierer, a graduate of Harvard Medical School, completed her internal medicine residency at the Massachusetts General Hospital and her hematology and medical oncology training at the Brigham and Women's Hospital and the Dana-Farber Cancer Institute. Dr. Bierer maintained a research laboratory in the Department of Pediatric Oncology at Dana-Farber Cancer Institute and was appointed Director of Pediatric Stem Cell Transplantation at Dana-Farber Cancer Institute and Children's Hospital in 1993. In 1997, she was named Chief of the Laboratory of Lymphocyte Biology at the National Heart, Lung and Blood Institute at the National Institutes of Health in Bethesda, MD. She served on the Scholars Committee of the Howard Hughes Medical Institute and on the Biomedical Research Training Program for Underrepresented Minorities at NHLBI, where she received the Director's Award in 1999. She returned to the Dana-Farber Cancer Institute in July 2002 as Vice President of Patient Safety and Director of the Center for Patient Safety. In 2003, Dr. Bierer moved to the Brigham and Women's Hospital to assume her current position. In addition, in 2006, Dr. Bierer established the Center for Faculty Development and Diversity at the Brigham and Women's Hospital and now serves as its first director.

Dr. Bierer's laboratory research interests include mechanisms of T cell activation and of immunosuppression, interests that complement her clinical commitment to hematology. In addition to her academic responsibilities, Dr. Bierer was elected to the Board of Directors of the Association for Accreditation of Human Research Protection Programs (AAHRPP), serving as its President from 2003-2007, and was on the Board of Directors of the Federation of American Societies for Experimental Biology (FASEB). She was a member of the Medical and Scientific Advisory Board and, later, the Board of Directors of ViaCell, Inc. She is on the editorial boards of a number of journals including *Current Protocols of Immunology*. She is currently a member of the AAMC-AAU Advisory Committee on Financial Conflicts of Interest in Clinical Research, on the National Academies' Committee on Science, Technology and the Law, and on the Secretary's Advisory Committee for Human Research Protections for which she serves as chair.

## MEMBERS

**Joseph Fraumeni, Jr.**, is a physician and cancer researcher. He received an A.B. from Harvard College, an M.D. from Duke University, and an M.Sc. in epidemiology from the Harvard School of Public Health. He completed his medical residency at Johns Hopkins Hospital and Memorial Sloan-Kettering Cancer Center, and received internal medicine board certification. He then joined the National Cancer Institute at the National Institutes of Health as a post-doctoral fellow, becoming founding Chief of the Environmental Epidemiology Branch in 1975, Director of the Epidemiology and Biostatistics Program in 1979, and currently Director of the Division of Cancer Epidemiology and Genetics since 1995.

Dr. Fraumeni has been recognized by numerous honors and awards for his research into the genetic and environmental determinants of cancer. Among them are the Abraham Lilienfeld Award from the American College of Epidemiology, John Snow Award from the American Public Health Association, James D. Bruce Award from the American College of Physicians, Nathan Davis Award from the American Medical Association, Charles S. Mott Prize (with F.P. Li) from the General Motors Cancer Research Foundation, Medal of Honor from the International

Agency for Research on Cancer, and Lifetime Achievement Award from the American Association for Cancer Research. Dr. Fraumeni is an elected member of the National Academy of Sciences, the Institute of Medicine, and the Association of American Physicians. He has over 800 scientific publications, including several books on the causes and prevention of cancer.

**Patricia Ganz** is Professor of Health Services in the School of Public Health and Professor of Medicine in the David Geffen School of Medicine at UCLA, and Vice Chair of the Department of Health Services. She teaches Health Care Practices and Variations and Ethical Issues in Public Health. Dr. Ganz received her M.D. from the University of California, Los Angeles, School of Medicine in 1973, and completed post-doctoral training in internal medicine and medical oncology at the UCLA Medical Center. She has been on the faculty of the School of Medicine since 1977, and joined the faculty of the School of Public Health in 1992. Dr. Ganz has devoted the past 25 years to the study of quality-of-life outcomes in cancer and other chronic diseases. She is a leader in the integration of quality-of-life assessment in clinical trials. She has conducted federally funded research for over two decades, and has led several large clinical intervention trials in breast cancer. Her current research focuses on the late effects of cancer treatment, and improving the quality of care for cancer survivors. In 2006, she received funding from the Lance Armstrong Foundation to establish the UCLA-LIVESTRONG Survivorship Center of Excellence, whose mission is to facilitate improvements in the quality of life and quality of care of cancer survivors in the Los Angeles region and wherever they may reside. Dr. Ganz is also the Director of the Division of Cancer Prevention and Control Research of the Jonsson Comprehensive Cancer Center at UCLA, and leads a large research group that applies the scientific disciplines of public health (epidemiology, health services, behavioral sciences, biostatistics) to research on the prevention, detection, treatment and supportive care of cancer. Dr. Ganz is Associate Editor of the *Journal of Clinical Oncology* and the *Journal of the National Cancer Institute* and is a member of the editorial board of the Cochrane Breast Cancer Group. In 1999 she was named an American Cancer Society Clinical Research Professor and in 2007 she became a member of the Institute of Medicine.

**Mikhail Gishizky**, Ph.D., has over 25 years experience in research and development within the academic, biotechnology and pharmaceutical industry settings where he led efforts in the development of revolutionary signal transduction inhibitor drugs for the treatment of cancers and other diseases. Dr. Gishizky has been instrumental in establishing two biotechnology companies (SUGEN, Entelos) whose technology is helping to bring the promise of personalized medicine to the patient's bedside. Most recently, as the Chief Scientific Officer at Entelos, Dr. Gishizky supervised the development of capabilities and scientific programs that employ computer simulation models to identify patient populations who would benefit most from new medicines and combination therapies. Use of these capabilities result in faster and more cost-effective drug development programs by helping drug developers predict patient responses prior to initiation of therapy, thus helping physicians optimize the beneficial outcome and minimize the risk to the patient. Dr. Gishizky has been a member of the Institute of Medicine Forum on Drug Discovery, Development and Translation since 2005. Earlier in his career, Dr. Gishizky held positions of increasing management responsibility at SUGEN, Pharmacia, and Pfizer (as Vice President and Research Zone Head), developing targeted therapies and signal transduction pathway analysis tools to identify patients most likely to respond to given therapies (e.g., Sutent, a leader in the class of signal transduction inhibitors marketed by Pfizer). Dr. Gishizky received his degree in Endocrinology at the University of California, San Francisco, where his work focused on defining the molecular mechanisms responsible for the development and progression of Diabetes Mellitus. Dr. Gishizky's postdoctoral training and academic work focused on cancer biology and hematopoietic cell development. His research led to the development of in vitro systems and an animal model for human Chronic Myeloid Leukemia that was instrumental in the development of Gleevec. Dr. Gishizky has published extensively in the areas of Diabetes Mellitus and oncology research. During his tenure within the biotech/pharma industry Dr. Gishizky has led research efforts across a broad range of therapeutic areas including oncology, immunology, inflammation, and CNS and metabolic diseases.

**Alberto Gutierrez**, Ph.D., is the Director of U.S. Food and Drug Administration's Office of In Vitro Diagnostic Device Evaluation and Safety. Dr. Gutierrez received a bachelor's degree from Haverford College, and master and doctorate degrees in Chemistry from Princeton University. Dr. Gutierrez has over 10 years of experience in research in the area of structural organic and organometallic chemistry. Dr. Gutierrez joined the FDA in 1992 as researcher and reviewer in FDA's Center for Biologics Evaluation and Research working on vaccine adjuvants and method development for determination of purity and structure of vaccine components. In 2000, he joined the Office of In Vitro Diagnostic Device Evaluation and Safety as a scientific reviewer, becoming a Team leader for Toxicology in 2003, Director of the Division of Chemistry and Toxicology Devices in 2005, Deputy Director of the Office of In Vitro Diagnostic Devices in 2007 and Director in 2009.

**Kathy Hudson** is Chief of Staff to the Director of the National Institutes of Health. She is the founder of the Genetics and Public Policy Center at Johns Hopkins University. At the time of the workshop, Dr. Hudson was Director of the Genetics and Public Policy Center and an Associate Professor in the Berman Bioethics Institute, Institute of Genetic Medicine, and the Department of Pediatrics at Johns Hopkins University. The Genetics and Public Policy Center was established in April 2002 with a grant from The Pew Charitable Trusts. Dr. Hudson founded the Center to fill an important niche in the science policy landscape and to focus exclusively on public policy issues raised by advances in human genetics. She led the Center's efforts to address legal, ethical, and policy issues related to human reproductive genetic technologies, genetic testing quality and oversight, and public engagement in genetic research. Prior to establishing the Center, Dr. Hudson served as Assistant Director of the National Human Genome Research Institute. She received her Ph.D. in molecular biology from the University of California, Berkeley.

**Muin Khoury** is the first director of the Centers for Disease Control and Prevention's Office of Public Health Genomics. The Office was formed in 1997 to assess the impact of advances in human genetics and the Human Genome Project on public health and disease prevention. As the nation's prevention agency, CDC's mission is to protect the health and safety of people, to provide credible information to enhance health decisions, and to promote health through strong partnerships. CDC's Office of Public Health Genomics serves as the national focus for integrating genomics into public health research and programs for disease prevention and health promotion. Dr. Khoury joined CDC as an Epidemic Intelligence Service Officer in 1980 in the Birth Defects and Genetic Diseases Branch, and as a medical epidemiologist in 1987. In

1990, he became Deputy Chief of the same Branch. In 1996, Dr. Khoury chaired a CDC-wide Task Force on Genetics and Disease Prevention and provided important leadership in outlining a plan delineating the future direction that CDC should take in this important area.

Dr. Khoury received his B.S. degree in Biology/Chemistry from the American University of Beirut, Lebanon and his medical degree and Pediatrics training from the same institution. He received a Ph.D. in Human Genetics/Genetic Epidemiology and training in Medical Genetics from Johns Hopkins University. Dr. Khoury is board certified in Medical Genetics.

Dr. Khoury has published extensively in the fields of genetic epidemiology and public health genetics. He is a member of many professional societies and serves on the editorial boards of several journals. Dr. Khoury also serves on several scientific, public health, and health policy national and international committees. He is an adjunct professor of Epidemiology at Emory's School of Public Health and an associate in the Department of Epidemiology at Johns Hopkins University Bloomberg School of Public Health.

**David Korn** became Vice Provost for Research of Harvard University in November 2008. He is also Professor of Pathology at Harvard Medical School. Prior to that he was senior vice president for biomedical and health sciences research at the Association of American Medical Colleges in Washington, DC, a position he assumed on September 1, 1997. Dr. Korn served as Carl and Elizabeth Naumann Professor and Dean of the Stanford University School of Medicine from October 1984 to April 1995, and as Vice President of Stanford University from January 1986 to April 1995. Before that he had served as Professor and Chairman of the Department of Pathology at Stanford, and Chief of the Pathology Service at the Stanford University Hospital, since June 1968. Dr. Korn has been Chairman of the Stanford University Committee on Research; President of the American Association of Pathologists (now the American Society for Investigative Pathology). Dr. Korn was a founder and Chairman of the Board of Directors of the California Transplant Donor Network, one of the nation's largest Organ Procurement Organizations. He is a member of the Institute of Medicine and a founder of the Clinical Research Roundtable. Dr. Korn served on the Boards of Directors of the Stanford University Hospital from October 1982 to April 1995, the Children's Hospital at Stanford from October 1984 to its closure, and the Lucile Salter Packard Children's Hospital at Stanford from October

1984 to April 1995. He was a member of the Board of Directors of the California Society of Pathologists from 1983-86. Since January 2010, Dr. Korn has been Co-chair of the National Academies' Committee on Science, Technology, and Law. He has been a member of the committee since its inception in 1998.

**Jonathan D. Moreno**, Ph.D., (IOM) is the David and Lyn Silfen University Professor of Ethics and Professor of Medical Ethics and of History and Sociology of Science at Penn. He holds a courtesy appointment as Professor of Philosophy. He is also a Senior Fellow at the Center for American Progress in Washington, DC, where he edits the magazine *Science Progress* (www.scienceprogress.org). He was a member of President Barack Obama's transition team for the Department of Health and Human Services. Dr. Moreno is an elected member of the Institute of Medicine/National Academy of Sciences and serves on numerous National Academies committees. In 2008 he was designated a National Associate of the National Research Council. He has served as a senior staff member for two presidential advisory commissions, and has given invited testimony for both houses of congress. He was an Andrew W. Mellon post doctoral fellow, holds an honorary doctorate from Hofstra University, and is a recipient of the Benjamin Rush Medal from the College of William and Mary Law School. Dr. Moreno has served as adviser to the Howard Hughes Medical Institute and the Bill and Melinda Gates Foundation, among many other organizations. Moreno is also a Faculty Affiliate of the Kennedy Institute of Ethics at Georgetown University and a Fellow of the Hastings Center and the New York Academy of Medicine. He is a past president of the American Society for Bioethics and Humanities. *Publisher's Weekly* said that his most recent book, *Science Next: Innovation for the Common Good* (2009), "brings hope into focus with reports of innovation that will enhance lives." His other books include *Mind Wars: Brain Research and National Defense* (2006), which the journal *Nature* called "fascinating and sometimes unsettling"; *Undue Risk: Secret State Experiments on Humans* (1999), described by the *The New York Times* as "an earnest and chilling account" and by the *Journal of the American Medical Association* as a "classic" in the literature on human experimentation; *Ethical Guidelines for Innovative Surgery (2006); Is There an Ethicist in the House?* (2005); *In the Wake of Terror: Medicine and Morality in a Time of Crisis* (2003); *Ethical and Regulatory Aspects of Clinical Research* (2003); *Deciding Together: Bioethics and Moral Consensus* (1995); *Ethics in Clinical Practice* (2000); and *Arguing*

*Euthanasia* (1995). Moreno has published more than 300 papers, reviews and book chapters, and is a member of several editorial boards.

## STAFF

**Anne-Marie Mazza** is Director of the Committee on Science, Technology and Law. She joined the National Academies in 1995. She has served as Senior Program Officer with both the Committee on Science, Engineering, and Public Policy and the Government-University-Industry Research Roundtable. In 1999 she was named the first director of the Committee on Science, Technology, and Law (CSTL), a newly created Program designed to foster communication and analysis among scientists, engineers, and members of the legal community. In 2007, she became the director of the Christine Mirzayan Science and Technology Graduate Policy Fellowship Program. Dr. Mazza has been the study director on numerous Academy reports including *Science and Security in a Post 9-11 World, 2007; Reaping the Benefits of Genomic and Proteomic Research, 2005*; *Intentional Human Dosing Studies for EPA Regulatory Purposes: Scientific and Ethical Issues, 2004; Ensuring the Quality of Data Disseminated by the Federal Government, 2003; The Age of Expert Testimony: Science in the Courtroom, 2002; Issues for Science and Engineering Researchers in the Digital Age, 2001; and Observations on the President's Fiscal Year 2000 Federal Science and Technology Budget, 1999.* Between October 1999 and October 2000, she divided her time between CSTL and the White House Office of Science and Technology Policy, where she served as a Senior Policy Analyst responsible for issues associated with the government-university research partnership. Before joining the National Academies, Dr. Mazza was a Senior Consultant with Resource Planning Corporation. She received a B.A., M.A., and Ph.D. from The George Washington University.

**Guruprasad Madhavan** was a Program Officer for the Committee on Science, Technology, and Law, and the Committee on Science, Engineering, and Public Policy at the National Academies until November 2010. He was formerly a Christine Mirzayan Science and Technology Policy Fellow with the National Academies' Board on Science, Technology, and Economic Policy. Madhavan received his B.E. (honors with distinction) in instrumentation and control engineering from the University of Madras (India), and M.S. in biomedical engineering from SUNY Stony Brook. Following his medical device industry experience as a research scientist at AFx, Inc. and Guidant Corporation in California, Madhavan completed an M.B.A. in leadership and health care management and a Ph.D. in biomedical engineering at SUNY Binghamton. His doctoral research was focused on non-invasive and non-pharmacologic neuromuscular stimulation for enhancing lower limb circulation. Among other honors, Madhavan was chosen as an outstanding young scientist to attend the 2008 World Economic Forum's Annual Meeting of the New Champions and as one of the 2009 *New Faces of Engineering* by the Engineers Week Foundation in the *USA Today*. He is also an elected member to the administrative council of the International Federation for Medical and Biological Engineering. Madhavan is a coeditor of *Career Development in Bioengineering and Biotechnology* (Springer) and *Pathological Altruism* (Oxford University Press).

**Steven Kendall** is Senior Program Associate for the Committee on Science, Technology, and Law. He is a Ph.D. candidate in the Department of the History of Art and Architecture at the University of California, Santa Barbara where he is completing a dissertation on 19th century British painting. Mr. Kendall received his M.A. in Victorian Art and Architecture at the University of London. Prior to joining The National Academies in 2007, he worked at the Smithsonian American Art Museum and The Huntington in San Marino, California.

**Mary Fraker** is a writer specializing in the intersection of emerging technologies and evolving public policy. She has written about financial institutions' roles and responsibilities under the USA PATRIOT Act; the disposition of wastewater treatment residuals, including siting issues and renewable-energy production; and encryption, electronic signatures and authenticating identities on the Internet. In the biomedical arena, she has written annual reports for, among others: Human Genome Sciences, 1993 (the company's first); Genetic Therapy Inc. (later acquired by Novartis), 1991 and 1992; and Life Technologies Inc., 1986 through 1990. From 1979 to 1991, Ms. Fraker held various marketing and communications positions at Life Technologies (formerly Bethesda Research Laboratories), writing journal advertisements, brochures and annual reports and managing the company's DNA sequencing workshop program. Between 1991 and 1994, she audited the open meetings of the Human Genome Project's ELSI (ethical, legal and social implications) Task Force and Insurance Working Group and served on the local (Washington, DC) steering committees of the Biotechnology Industry

Organization (BIO) and the Association of Biotechnology Companies (one of BIO's two predecessor organizations). Ms. Fraker holds bachelor and masters degrees, respectively, from Tufts University and Canada's York University.

**Adam C. Berger, Ph.D.** is a Program Officer at the Board on Health Sciences Policy at the Institute of Medicine where his primary interest focuses on policy issues relating to translational medicine. Dr. Berger received his doctorate from Emory University in the Biochemistry, Cell and Developmental Biology Program studying the fatal childhood neurodegenerative disorder, Niemann Pick Disease Type C, and additionally performing genome wide screening for novel cellular targets of anti-cancer and anti-malarial drug compounds. Following a post-doctoral fellowship at the National Cancer Institute studying the immunological impact of altered cholesterol homeostasis, Dr. Berger joined the Institute of Medicine where he is currently directing the Roundtable on Translating Genomic-Based Research for Health. He is the recipient of the National Institutes of Health Fellows Award for Research Excellence and a Ruth L. Kirschstein National Research Service Award. Dr. Berger received his B.S. from The Ohio State University where he majored in Molecular Genetics.

**Sally Cluchey** (formerly Robinson) is a Research Associate with the Engelberg Center for Health Care Reform at The Brookings Institution. She is responsible for working with multiple stakeholders to develop the infrastructure, methods, and governance structure necessary to conduct active medical product safety surveillance and comparative effectiveness research.

Prior to joining Brookings she was a program officer with the Institute of Medicine (IOM) of the National Academy of Sciences. At IOM, Ms. Cluchey staffed multiple projects including the Committee on Comparative Effectiveness Research Prioritization, where she helped to write the report *Initial Prioritizes for Comparative Effectiveness Research*, and served for two years on the Forum on Drug Discovery, Development, and Translation. While working on the Forum, Ms. Cluchey was responsible for several key IOM initiatives involving the science of drug safety, FDA policy, funding models for drug development, and improving the clinical research process.

Prior to joining the IOM in 2006, she worked for the Walter Reed Army Institute of Research's malaria vaccine development program, where she managed the manufacture, preclinical, and Phase I development of multiple vaccine candidates, and coordinated regulatory submissions. Ms. Cluchey holds a Master of Science in Biomedical Science and Regulatory Compliance from Hood College and a Bachelor of Arts from Kenyon College.

**Lyla M. Hernandez** (until March 2010) was a Senior Program Officer with the Institute of Medicine for the past thirteen years. During that time she has been study director for projects related to Gulf War veterans' health, public health, complementary and alternative medicine, genomics, and health status indicators. Reports from her most recently completed studies include *Genes, Behavior, and the Social Environment: Moving Beyond the Nature/Nurture Debate*, *Training Physicians for Public Health Careers*, and *State of the USA Health Indicators*. She is currently Staff Director of the Roundtable on Translating Genomic-Based Research for Health, Staff Director of the Roundtable on Health Literacy, and Study Director for the Committee on a National Surveillance System for Cardiovascular and Other Select Chronic Diseases. Prior to joining the IOM, Lyla was Director of the Pharmacy Intelligence Center of the American Pharmaceutical Association where she was responsible for identifying and conducting analyses of health care issues, national programs, and policies that affect pharmacy. She also served as Executive Director of the American Medical Peer Review Organization, the national trade association for organizations evaluating the utilization and quality of medical care. Lyla received her Masters of Public Health degree from the University of California at Berkeley and her Bachelor of Science in Education from the University of Illinois at Urbana.

**Sharon Murphy** joined the Institute of Medicine as a Scholar-in-Residence in October 2008, coming to DC from Texas where she was the inaugural Director of the Greehey Children's Cancer Research Institute and Professor of Pediatrics at the University of Texas Health Science Center at San Antonio from 2002 to 2008. From 1988 to 2002, Dr. Murphy was Chief of the Division of Hematology/Oncology at Children's Memorial Hospital in Chicago and Professor of Pediatrics at Northwestern University School of Medicine where she also led the program in pediatric oncology at the Robert H. Lurie Cancer Center. From 1974 to 1988, Dr. Murphy was on the faculty at St. Jude Children's Research Hospital in Memphis. A pediatric oncologist and clinical cancer researcher, Dr. Murphy has devoted the past thirty-five years to improving cure rates for childhood cancer, particularly childhood lymphomas and leukemias.

She was chair of the Pediatric Oncology Group from 1993 to 2001. She has been recognized for her achievements by the Association of Community Cancer Centers (2001), the Distinguished Service Award for Scientific Leadership from the American Society of Clinical Oncology (2005) and the Distinguished Career Award from the American Society of Pediatric Hematology and Oncology (2009). The author of more than 220 original articles, reviews, and book chapters, Dr. Murphy has also served on numerous editorial boards, including *Cancer Research*, *Clinical Cancer Research*, and the *Journal of Clinical Oncology*. She has been a member of the Boards of Directors of the American Cancer Society, the American Association of Cancer Research, the American Society of Hematology, and the American Society of Clinical Oncology, and has been an advisor to NCI and FDA. She earned her bachelor's of science degree from the University of Wisconsin (1965) and her medical degree, cum laude, from Harvard Medical School (1969). She completed postdoctoral training in pediatrics at the University of Colorado (1969-71) and in pediatric hematology and oncology at the University of Pennsylvania (1971-73).

**Ann Reid** was a Senior Program Officer for the Board on Life Sciences at the National Research Council until December 2009. She served as study director for a diverse set of reports including: *The Ecological Impacts of Climate Change*; *The Role of Theory in Advancing 21st-Century Biology*; *The New Science of Metagenomics*; *Exploring the Role of Antiviral Drugs in Eradicating Polio*; and *Treating Infectious Diseases in a Microbial World*. From 1989 to 2004 she was a research biologist at the Armed Forces Institute of Pathology where she applied the techniques of molecular biology to archival tissue samples. From 1995 to 2004 her research focused on the isolation and sequencing of the 1918 pandemic influenza virus from archived autopsy samples and lung samples from an Inuit victim who had been buried in permafrost. The genomic sequence of the virus was completed in 2004. The sequence has been used to try to determine why the 1918 epidemic was so severe, where the pandemic strain came from, and to test the effectiveness of influenza vaccines and antiviral drugs. Before turning to science she was a political analyst for the Japanese Embassy in Washington, DC, and the Organization for Economic Cooperation and Development in Paris. She holds a B.A. in environmental science from Simon's Rock College, an M.A. from the Johns Hopkins School of Advanced International Studies and has published more than 30 papers and reviews, and 7 book chapters.

**Adam Schickedanz** was a Christine Mirzayan Science and Technology Policy Fellow with the Institute of Medicine's Board on Health Care Services in winter of 2009 and served as a member of the IOM staff for nearly a year. Mr. Schickedanz received his baccalaureate degree from Washington University in St. Louis and his M.D. from the University of California, San Francisco, in May of 2010. At UCSF, he developed a clinical focus in urban underserved patient care while advancing research interests in professionalism and cultural competency in medical education, novel approaches to clinician-patient communication in medical decision making (particularly at the end of life), and the intersections of education and health. At IOM, he contributed projects on value in cancer care, health consequences for the uninsured, continuing medical education, and health care data collection practices to improve quality of care and reduce health disparities. Mr. Schickedanz will pursue residency training in pediatrics.

# Appendix F

# Workshop Agenda

**Direct-to-Consumer Genetic Testing:**
**A Cross-Academies Workshop**
The National Academies
Washington, DC
August 31-September 1, 2009

Moderators:    **Frederick R. Anderson**, Jr., Partner, McKenna, Long & Aldridge LLP

**Barbara E. Bierer**, Professor of Medicine, Harvard Medical School and Senior Vice President, Research, Brigham and Women's Hospital

**August 31, 2009**

7:30            Breakfast

8:00-10:40      *Session 1: Overview of Research on DTC Genetic Testing and Its Trajectory*

Direct-to-consumer genetic testing represents a $730 million global market, with projected growth of 20 percent annually. While many direct-to-consumer genetic tests assess risk for illnesses with strong genetic heritability and raise concerns over adequate counseling and appropriate outlets for such information, still other genetic tests to guide risk management for diseases with much smaller genetic components, or no clear genetic basis at all, have rapidly emerged and present new dilemmas for consumers and health care providers alike. With the costs of genetic analyses falling rapidly and entrepreneurs finding more and more creative uses for these technologies and the test results they produce, the future of genetic testing is being ushered in with both the hope that its tremendous promise will be realized and concern over the accompanying cultural, professional, and regulatory challenges to be faced.

Issues to Address: Direct-to-consumer genetic tests have uncertain analytical and clinical validity, and questionable clinical utility. What exactly can one tell based on these tests? What can't one tell? This will have implications for the testing companies' claims. What types of genetic testing will become available over the next five to ten years? What will the future market look like?

| | |
|---|---|
| 8:00 | Introduction to the Scope of the Workshop |
| | **David Korn**, Vice Provost for Research, Harvard University |
| 8:20 | Drivers of Innovation: The Human Genome Project, Microarrays, the HapMap and the $1,000 Genome |
| | **Alan Guttmacher**, Acting Director, National Human Genome Research Institute, National Institutes of Health |
| 8:40 | Discussion |
| 9:00 | Direct-to-Consumer Genetic Testing: History and Scientific Foundation |
| | **Muin Khoury**, Director, Office of Public Health Genomics, Centers for Disease Control and Prevention |
| 9:20 | Discussion |
| 9:40 | Evolution of Direct-to-Consumer Genetic Testing: Present and Future Markets |

**K. David Becker**, Chief Scientific Officer, Pathway Genomics Corporation

10:00      Discussion

10:20      Break

10:40-12:45   *Session 2: The Regulatory Framework*

With the implementation of the Clinical Laboratory Improvement Act (CLIA) predating major advances in genetics and the FDA only able to regulate genetic test kits, the vast majority of lab-derived genetic testing operates with sparse regulatory oversight relative to other laboratory tests of comparable capacity to explain and predict health and disease. In addition, no claims by direct-to-consumer genetic testing companies have been challenged by the FTC to date. As the technology advances at a dizzying clip and consumer interest continues to grow, the lagging federal, state, professional and consumer regulatory entities will need to consider how best to ensure valid tests and accurate advertising without stymieing innovations that promise what could be the next medical paradigm shift—or is this just more unregulated hype?

Issues to Address: Differentiating regulatory issues for DTC testing vs. genetic testing generally. DTC-specific regulatory issues include examining whether oversight of advertising/claims is adequate. Are the claims verifiable? Spell out the roles of various agencies in oversight, state, and federal roles. What is the impact of regulatory uncertainty on DTC companies? What are the codes of professional conduct for informed consent, analysis, and disclosure? Is it possible to create safeguards without hindering rapid technological advances? If testing procedures aren't "approved" can they be quality assured?

| 10:40 | Existing oversight of genetic testing in the U.S. and U.K. |
|---|---|

**Andrea Ferreira-Gonzalez**, Professor of Pathology, Virginia Commonwealth University and Director, The Molecular Diagnostics Laboratory, Virginia Commonwealth University Health System

**Timothy Aitman**, Professor of Clinical and Molecular Genetics, Division of Clinical Sciences, Imperial College London

| 11:20 | Discussion |
|---|---|

| 11:55 | Monitoring direct-to-consumer genetic testing |
|---|---|

**Gregory Kutz**, Managing Director, Forensic Audits and Special Investigations, Government Accountability Office

**Sandra Soo-Jin Lee**, Senior Research Scholar, Stanford Center for Biomedical Ethics

| 12:30 | Discussion |
|---|---|

| 1:00 | Lunch |
|---|---|

| 1:45-3:20 | *Session 3: Shared Genes and Emerging Issues in Privacy* |
|---|---|

Genetic information has implications for the health and well-being of others beyond the individual whose DNA was sequenced, simultaneously suggesting the need to protect the sequenced individual from unethical treatment based on undesirable sequence and the potential responsibility to inform those (blood relatives, perhaps others) with a stake in the individual's genetics. The danger of untoward consequences of public genetic data is only further enhanced by the popular notion—correct in some cases but overly simplistic in many others—that genes are a biological "blueprint" by which attributes from shoe size to temperament are determined. While the Genetic Information Nondiscrimination Act forbids unequal treatment based on one's genetics in many circumstances, it seems naïve to suppose that exposing one's genetic frailties wouldn't pose considerable social risk.

<u>Issues to Address</u>: How to balance the desire for self-awareness among consumers that is driving this market against the need to protect privacy? What are the risks and benefits for family members of users of these tests? For public figures? For the legal system? Who owns an individual's genomic data? Discrimination issues and the effectiveness of GINA. Social networks based on direct-to-consumer genetic testing results.

| | |
|---|---|
| 1:40 | Existing Structures for Privacy and Nondiscrimination Protections: Beyond the Genetic Information Nondiscrimination Act |
| | **Susannah Baruch**, Policy Director, Generations Ahead and Policy Analyst, Genetics and Public Policy Center, Johns Hopkins University |
| 2:00 | Discussion |
| 2:20 | Genetic Identity and Community |
| | **Scott Woodward**, Director, Sorenson Molecular Genealogy Foundation |
| 2:40 | Discussion |
| 3:00 | Break |
| 3:20-4:30 | *Session 4: DTC Genetic Testing Companies and Research* |

Direct-to-consumer genetic testing companies have already taken steps to use the rich data their customers provide them in research to improve their products, to offer new services, and even to benefit the broader research community. It stands to reason that protections for consumers-turned-research-subjects should be equivalent to those for human participants in academic genetics research, but no systematized mechanism for ensuring these protections currently exists. In addition, who should the results of the research benefit? Who owns this information and, by taking on a research role that could serve the public good, do direct-to-consumer testing companies assume an ethical responsibility to ensure that the public benefits?

Issues to Address: Who retains ownership of genetic information when companies use their testing data for research? (IP) Who should have access to the results? Who should be allowed to benefit from the advances as a result of this research? Will genetic information and the research it spurs become a private commodity?

| | |
|---|---|
| 3:20 | Direct-to-Consumer Genetic Testing Companies as Research Entities: Disclosure, Intellectual Property, and Shared Advances |
| | **Elissa Levin**, Director, Genetic Counseling Program, Navigenics, Inc. |
| 3:40 | Discussion |
| 4:30 | Adjourn |

Dinner

**September 1, 2009**

| | |
|---|---|
| 7:30 | Breakfast |

8:15        The FDA and the Regulation of Direct-to-Consumer Genetic Testing

            **Courtney Harper**, Acting Director of the Division of Chemistry and Toxicology Devices, Office of In Vitro Diagnostic Device Evaluation and Safety, Center for Devices and Radiological Health, U.S. Food and Drug Administration

8:40        Q&A

9:00-12:00  *Session 5: The Impact of DTC Genetic Tests on the Medical System*

            If the medical system is no longer required to mediate genetic testing, how will the system cope with losing oversight (and reimbursement) of these services while retaining the full responsibility of caring for patients the services affect?

            Issues to Address: Can we model the cost to the medical system of DTC genetic testing? Reimbursement and DTC genetic testing—are insurance companies involved? Do they have a role? How can providers navigate DTC testing and results for patients in the clinic? How do consumers react to DTC testing information, and what is the impact on their health behavior?

9:00        What Are the Costs and Benefits to the Health Care System?

            **Kathryn Phillips**, Professor of Health Economics and Health Services Research, University of California, San Francisco

9:20        Discussion

9:40            Knowledge of DTC Genetic Testing Among the Public
                and Health Professionals Public Understanding

                **Katrina Goddard**, Senior Investigator, Kaiser Perman-
                ente Center for Health Research

10:00           Understanding Among Health Professionals

                **Joseph McInerney**, Executive Director, National Coali-
                tion for Health Professional Education in Genetics

10:20           Discussion

10:40           Cooperation or Competition—How Do Health Care
                and DTC Genetic Testing Coexist?

                **Patricia Ganz**, Professor of Health Services, School of
                Public Health and Professor of Medicine, David Geffen
                School of Medicine, University of California, Los Angeles

11:00           Discussion

11:20           The Impact of Direct-to-Consumer Genetic Testing on
                Public Health

                **Harvey Fineberg**, President, Institute of Medicine

11:40           Summary Discussion

12:00           Lunch / Adjourn

# Appendix G

# Workshop Participants

**Direct-to-Consumer Genetic Testing:**
**A Cross-Academies Workshop**
The National Academies
Washington, DC
August 31-September 1, 2009

Tanya Agurs-Collins, National Institutes of Health National Cancer
    Institute
Vincent Allen, Jr., National Institutes of Health National Human
    Genome Research Institute
Richard Apostol
Neeraj Arora, National Institutes of Health National Cancer Institute
Suresh Arya, National Institutes of Health
Ronald Bailey, Reason Magazine
Judith Benkendorf, American College of Medical Genetics
Barbara Bernhardt, Penn Center for the Integration of Genetic Health
    Care Technologies
Barbara Biesecker, National Institutes of Health
Laura Bishop, Kennedy Institute of Ethics, Georgetown University
Juli Bollinger, Genetics and Public Policy Center
Vence Bonham, National Institutes of Health National Human
    Genome Research Institute
Malorye Branca, Nature Biotechnology
Stan Brull, Center for Bioethics, University of Pennsylvania
Kee Chan, Boston University
Ashadeep Chandrareddy, Tufts Medical Center
Ralph Coates, Centers for Disease Control
Jennifer Costello, U.S. Government Accountability Office
Arkendra De, U.S. Food and Drug Administration
Maria Della Rocca, Genetic and Rare Diseases Information Center
Tania Diaz

Jennifer Dolan, University of Pennsylvania Health System
Britt Erickson, Chemical & Engineering News
Cathy Fomous, National Institutes of Health Office of Biotechnology
    Activities
Keolu Fox, National Institutes of Health
Santhi Ganesh, National Institutes of Health
Susan Hahn, University of Miami
Patrick Hanaway, Genova Diagnostics
Karen Hanson, U.S. Food and Drug Administration
Sarah Harding, National Institutes of Health National Human
    Genome Research Institute
Emily Harris, National Institutes of Health National Institute of
    Dental and Craniofacial Research
Erin Heath, American Association for the Advancement of Science
Laura Hercher, Sarah Lawrence College Joan H. Marks Human
    Genetics Program
Christine Hodakievic, U.S. Government Accountability Office
Gillian Hooker, Georgetown University / Lombardi Comprehensive
    Cancer Center
Sandra Howard, U.S. Department of Health and Human Services
    Office of the Assistant Secretary for Planning and Evaluation
Jordanna Joaquina, AccessDNA
Sarah Kalia, National Institutes of Health Graduate Partnership
    Program
Pei Koay, Center for Contemporary History and Policy, Chemical
    Heritage Foundation
Kenneth Kornman, Interleukin Genetics
Vinayak Kottoor, Johns Hopkins University
Alanna Kulchak, Rahm Kaiser Permanente
Marcie Lambrix, Case Western Reserve University
Jane Lee, National Institutes of Health / Johns Hopkins University
Jennifer Leib, HealthFutures, LLC
Marcella McSorley, Global Clinical Solutions, Inc.
Radha Menon, U.S. Food and Drug Administration
Richard Merrill, University of Virginia
Isis Mikhail, National Institutes of Health National Cancer Institute
Francisco Murillo, U.S. Food and Drug Administration
Melanie Myers, University of Cincinnati and Cincinnati Children's
    Hospital Medical Center

Vivian Ota Wang, National Institutes of Health National
    Nanotechnology Coordination Office
Belen Pappa
Paul Partha, Interleukin Genetics
Andre Pilon, National Institutes of Health National Human Genome
    Research Institute
Susan Poland, Kennedy Institute of Ethics, Georgetown University
Tabitha Powledge, National Association of Science Writers
Andrew Pratt, Center for American Progress
Edward Ramos, National Institutes of Health
Erin Ramos, National Institutes of Health National Human Genome
    Research Institute
John Richardson, National Society of Genetic Counselors
Roslyn Rodriguez, University of the Philippines, Philippine General
    Hospital
Michael Rugnetta, Center for American Progress
Leah Sansbury, National Institutes of Health National Cancer Institute
Amanda Sarata, Congressional Research Service
Sheri Schully, National Institutes of Health National Cancer Institute
Joan Scott, Genetics and Public Policy Center
Mona Shah, American Cancer Society Cancer Action Network
Fay Shamanski, College of American Pathologists
Celia Sharp
Artemis Simopoulos, The Center for Genetics Nutrition and Health
Amanda Singleton, National Institutes of Health
Jennifer Smith, FDA Week
Pothur Srinivas, National Institutes of Health
Rebecca Sutphen, Moffitt Cancer Center at the University of South Florida
Melanie Swan, MS Futures Group
Zivana Tezak, Office of In Vitro Diagnostic Device Evaluation and
    Safety, Center for Devices and Radiological Health, U.S. Food and
    Drug Administration
Christopher Wade, National Institutes of Health National Human
    Genome Research Institute
Lan-Hsiang Wang, National Institutes of Health National Heart, Lung
    and Blood Institute
Kay Wanke, National Institutes of Health
Edward Winstead, National Institutes of Health National Cancer Institute
Emily Wold, U.S. Government Accountability Office
Kristi Zonno, Genetic Alliance

# Appendix H

# Membership of Oversight Boards and Committees

## COMMITTEE ON SCIENCE, TECHNOLOGY, AND LAW

**DONALD KENNEDY** (NAS/IOM), (Co-Chair), President Emeritus and Bing Professor of Environmental Science Emeritus, Stanford University; Emeritus Editor-in-Chief, *Science*

**RICHARD A. MERRILL** (IOM), (Co-Chair), Daniel Caplin Professor of Law Emeritus, University of Virginia Law School

**FREDERICK R. ANDERSON, JR.**, Partner, McKenna, Long & Aldridge LLP

**ARTHUR I. BIENENSTOCK**, Special Assistant to the President for SLAC and Federal Research Policy, Stanford University

**BARBARA E. BIERER**, Senior Vice President for Research, Brigham and Women's Hospital

**ELIZABETH H. BLACKBURN** (NAS/IOM), Morris Herzstein Professor of Biology and Physiology, Department of Biochemistry and Biophysics, University of California, San Francisco

**JOE S. CECIL**, Project Director, Program on Scientific and Technical Evidence, Federal Judicial Center

**RICHARD F. CELESTE**, President, Colorado College

**JOEL E. COHEN** (NAS), Abby Rockefeller Mauzé Professor and Head, Laboratory of Populations, The Rockefeller University and Columbia University

**ROCHELLE COOPER DREYFUSS**, Pauline Newman Professor of Law and Director, Engelberg Center on Innovation Law and Policy, New York University School of Law

**ALICE P. GAST** (NAE), President, Lehigh University

**LAWRENCE O. GOSTIN** (IOM), Associate Dean for Research and Academic Programs, Linda D. and Timothy J. O'Neill Professor

of Global Health Law, and Faculty Director, O'Neill Institute
for National and Global Health Law, Georgetown University;
Professor of Public Health, The Johns Hopkins University;
Director of the Center for Law & the Public's Health, The Johns
Hopkins and Georgetown Universities

**GARY W. HART**, Wirth Chair Professor, School of Public Affairs,
University of Colorado, Denver

**BENJAMIN W. HEINEMAN, JR.**, Senior Fellow, Harvard Law
School and Harvard Kennedy School of Government

**DAVID BROCK HORNBY**, Judge, U.S. District Court, District of
Maine

**DAVID KORN** (IOM), Chief Scientific Officer, Association of
American Medical Colleges

**RICHARD A. MESERVE** (NAE), President, Carnegie Institution for
Science and Senior Of Counsel, Covington & Burling, LLP

**DUNCAN T. MOORE** (NAE), Professor, The Institute of Optics,
University of Rochester

**ALAN B. MORRISON**, Attorney, Fail Elections Legal Network

**HARRIET RABB**, Vice President and General Counsel, Rockefeller
University

**PAUL D. RHEINGOLD**, Senior Partner, Rheingold, Valet,
Rheingold, Shkolnik & McCartney LLP

**BARBARA ROTHSTEIN**, Director, Federal Judicial Center

**JONATHAN M. SAMET** (IOM), Professor and Chairman,
Department of Epidemiology, The Johns Hopkins Bloomberg
School of Public Health

**DAVID S. TATEL**, Judge, U.S. Court of Appeals for the District of
Columbia


**Staff**

**ANNE-MARIE MAZZA**, Director
**GURUPRASAD MADHAVAN,** Program Officer (until November
2010)
**STEVEN KENDALL**, Senior Program Associate

# BOARD ON LIFE SCIENCES

**KEITH R. YAMAMOTO** (NAS/IOM), (Chair), Professor of Cellular and Molecular Pharmacology and Executive Vice Dean, School of Medicine, University of California, San Francisco

**ANN M. ARVIN** (IOM), Lucile Packard Professor in Pediatrics and Microbiology & Immunology and Vice Provost and Dean of Research, Stanford University School of Medicine

**BONNIE L. BASSLER** (NAS), Howard Hughes Medical Institute Investigator and Squibb Professor of Molecular Biology, Princeton University

**VICKI L. CHANDLER** (NAS), Chief Program Officer for Science, Gordon and Betty Moore Foundation and Regents' Professor, Departments of Plant Sciences and Molecular & Cellular Biology, University of Arizona

**SEAN EDDY**, Group Leader, Janelia Farm Research Campus, Howard Hughes Medical Institute

**MARK D. FITZSIMMONS,** Associate Director, MacArthur Fellows Program, John D. and Catherine T. MacArthur Foundation

**DAVID R. FRANZ**, Vice President and Chief Biological Scientist, Midwest Research Institute and Senior Advisor, Office of the Assistant to the Secretary of Defense for Nuclear, Chemical and Biological Defense Programs

**LOUIS J. GROSS,** James R. Cox Professor in Ecology and Evolutionary Biology and Mathematics and Director, The Institute for Environmental Modeling, University of Tennessee; and Director, National Institute for Mathematical and Biological Synthesis

**JO HANDELSMAN,** Howard Hughes Medical Institute Professor and Chair, Department of Bacteriology, University of Wisconsin–Madison

**CATO T. LAURENCIN** (IOM), Van Dusen Endowed Chair in Academic Medicine, Distinguished Professor of Orthopaedic Surgery, and Vice President for Health Affairs, University of Connecticut Health Center; and Dean, University of Connecticut School of Medicine, University of Connecticut Health Center

**JONATHAN D. MORENO** (IOM), David and Lyn Silfen University Professor of Ethics and Professor of Medical Ethics and of History and Sociology of Science, University of Pennsylvania; and Senior Fellow, Center for American Progress

**ROBERT M. NEREM** (NAE/IOM), Robert H. Petit Distinguished Chair for Engineering in Medicine and Institute Professor and Director, Parker H. Petit Institute for Bioengineering and Bioscience; and Director, Georgia Tech/Emory Center for the Engineering of Living Tissue, Georgia Institute of Technology

**CAMILLE PARMESAN,** Associate Professor of Integrative Biology, University of Texas at Austin

**MURIEL E. POSTON**, Dean of the Faculty and Professor of Biology, Skidmore College

**ALISON G. POWER,** Dean of the Graduate School and Professor of Ecology and Evolutionary Biology and of Science & Technology Studies, Cornell University

**BRUCE W. STILLMAN** (NAS), President and Chief Executive Officer, Cold Spring Harbor Laboratory

**CYNTHIA WOLBERGER**, Howard Hughes Medical Institute Investigator and Professor of Biophysics and Biophysical Chemistry, Johns Hopkins University School of Medicine

**MARY WOOLLEY** (IOM), President and Chief Executive Officer, Research!America

**Staff**

**FRANCES E. SHARPLES,** Director
**JO L. HUSBANDS**, Scholar/Senior Project Director
**ADAM P. FAGEN,** Senior Program Officer
**ANN H. REID,** Senior Program Officer
**MARILEE K. SHELTON-DAVENPORT,** Senior Program Officer
**INDIA HOOK-BARNARD**, Program Officer
**ANNA FARRAR,** Financial Associate
**AMANDA P. CLINE,** Senior Program Assistant (until May 2010)
**REBECCA L. WALTER**, Senior Program Assistant
**CARL-GUSTAV ANDERSON**, Program Assistant

# FORUM ON DRUG DISCOVERY, DEVELOPMENT, AND TRANSLATION

**GAIL CASSELL** (IOM), (Co-Chair), Vice President, Scientific Affairs and Distinguished Lilly Research Scholar for Infectious Diseases, Eli Lilly and Company

**JEFFREY DRAZEN** (IOM), (Co-Chair), Editor-in-Chief, *New England Journal of Medicine*

**BARBARA ALVING**, Director, National Center for Research Resources, National Institutes of Health

**LESLIE BENET** (IOM), Professor, Department of Biopharmaceutical Sciences, School of Pharmacy, University of California, San Francisco

**ANN BONHAM**, Chief Scientific Officer, Association of American Medical Colleges

**CATHERINE BONUCCELLI**, Vice President, Development Projects, Symbicort, AstraZeneca Pharmaceuticals

**LINDA BRADY**, Director, Division of Neuroscience and Basic Behavioral Science, National Institute of Mental Health

**ROBERT CALIFF**, Director, Duke Translational Medicine Institute, Professor of Medicine and Vice Chancellor for Clinical and Translational Research, Duke University Medical Center

**SCOTT CAMPBELL**, National Vice President, American Diabetes Association

**C. THOMAS CASKEY** (NAS/IOM), Chief Operating Officer and Director/CEO, University of Texas HSC at Houston

**PETER CORR**, Co-Founder and General Partner, Celtic Therapeutics LLLP

**JAMES DOROSHOW**, Director, Division of Cancer Treatment and Diagnosis, National Cancer Institute

**PAUL EISENBERG**, Senior Vice President, Global Regulatory Affairs and Safety, Amgen Inc.

**GARY FILERMAN**, Senior Vice President and Chairman, Health Management and Policy Group, ATLAS Research

**GARRET FITZGERALD** (IOM), Professor of Medicine, Professor and Chair of Pharmacology, Department of Pharmacology, University of Pennsylvania School of Medicine

**ELAINE GALLIN**, Director, Medical Research Programs, Doris Duke Charitable Foundation

**STEVEN GALSON**, Senior Vice President, Science Operations International Corporation

**MIKHAIL GISHIZKY**, Chief Scientific Officer, Entelos, Inc.

**HARRY GREENBERG**, Senior Associate Dean for Research, Joseph D. Grant Professor of Medicine and Microbiology & Immunology, Stanford University School of Medicine

**STEPHEN GROFT**, Director, Office of Rare Disease Research, National Institutes of Health

**PETER HONIG**, Former Executive Vice President for Worldwide Regulatory Affairs and Product Safety within Development, Merck Research Laboratories

**ANNALISA JENKINS**, Senior Vice President Global Medical, Bristol Myers Squibb

**MICHAEL KATZ** (IOM), Senior Vice President for Research and Global Programs, March of Dimes Foundation

**JACK KEENE**, James B. Duke Professor, Molecular Genetics & Microbiology, Duke University Medical Center

**RONALD KRALL**, Senior Vice President and Chief Medical Officer, GlaxoSmithKline

**FREDA LEWIS-HALL**, Chief Medical Officer, SVP, Pfizer

**WILLIAM MATTHEW**, Director of Translational Research, NINDS National Institutes of Health

**MARK MCCLELLAN** (IOM), Director, Engelberg Center for Health Care Reform, The Brookings Institution

**CAROL MIMURA**, Assistant Vice Chancellor for Intellectual Property & Industry Research Alliances (IPIRA), University of California, Berkeley

**JOHN ORLOFF**, Senior Vice President, US Medical & Drug Regulatory Affairs, Novartis Pharmaceuticals Corporation

**AMY PATTERSON**, Acting Director, Office of Science Policy, National Institutes of Health

**JANET SHOEMAKER**, Director, Office of Public Affairs, American Society for Microbiology

**LANA SKIRBOLL**, Acting Director, Division of Program Coordination, Planning, and Strategic Initiatives, National Institutes of Health

**NANCY SUNG**, Senior Program Officer, Burroughs-Wellcome Fund

**JORGE TAVEL**, Deputy Director, Division of Clinical Research, National Institute of Allergy and Infectious Diseases

**JANET TOBIAS**, President Sierra/Tango Productions, Ikana Media

**JOANNE WALDSTREICHER**, Head of Global Development, CNS/IM Franchise

**JANET WOODCOCK**, Deputy Commissioner and Chief Medical Officer, U.S. Food and Drug Administration

**RAYMOND WOOSLEY**, President and CEO, The Critical Path Institute

**Staff**

**ROBERT GIFFIN**, Forum Director (until February 2010)

**SALLY CLUCHEY**, Program Officer (until October 2009)

## ROUNDTABLE ON TRANSLATING GENOMIC-BASED RESEARCH FOR HEALTH

**WYLIE BURKE**, (IOM), (Chair), Professor and Chair, Department of Bioethics and Humanities, University of Washington

**BRUCE BLUMBERG,** Co-Chief of Medical Genetics, Kaiser-Permanente, Oakland, and Institutional Director of Graduate Medical Education, Northern California Kaiser Permanente

**DENISE E. BONDS,** Medicial Director, Division of Prevention & Population Sciences, National Heart, Lung, and Blood Institute

**C. THOMAS CASKEY** (NAS/IOM), Director and Chief Executive Officer, The George & Cynthia Mitchell Distinguished Chair in Neurosciences and Executive Vice President of Molecular Medicine and Genetics, Brown Foundation Institute of Molecular Medicine, University of Texas Health Science Center at Houston

**STEPHEN ECK,** Vice President, Translational Medicine & Pharmacogenomics, Eli Lilly and Company

**ANDREW N. FREEDMAN,** Genetic Epidemiologist, Risk Factor Monitoring and Methods Branch, Division of Cancer Control and Population Sciences, National Cancer Institute

**GEOFFREY GINSBURG,** Director, Center for Genomic Medicine, Institute for Genomic Sciences & Policy, Duke University

**R. RODNEY HOWELL,** Special Assistant to the Director, National Institute of Child Health and Human Development

**KATHY HUDSON**, Director, Genetics and Public Policy Center, Johns Hopkins Berman Institute of Bioethics (until July 2009)

**SHARON KARDIA**, Director, Public Health Genetic Programs Associate Professor, Department of Epidemiology, University of Michigan, School of Public Health

**MOHAMED KHAN,** Associate Director of Translational Research Department of Radiation Medicine, Roswell Park Cancer Institute

**MUIN KHOURY,** Director, National Office of Public Health Genomics, Centers for Disease Control and Prevention

**ALLAN KORN,** Chief Medical Officer, Senior Vice President Clinical Affairs, BlueCross/BlueShield Association

**DEBRA LEONARD,** Professor and Vice Chair of Laboratory Medicine Director of the Clinical Laboratories, Weill Cornell Medical Center

**MICHELE LLOYD-PURYEAR,** Chief, Genetic Services Branch, Health Resources and Services Administration

**ELIZABETH MANSFIELD,** Director of Personalized Medicine, Office of In Vitro Diagnostic Device Evaluation and Safety, Center for Devices and Radiological Health, U.S. Food and Drug Administration

**GARRY NEIL,** Corporate Vice President, Corporate Office of Science and Technology (COSAT), Johnson & Johnson

**ROBERT L. NUSSBAUM** (IOM), Chief, Division of Medical Genetics, Department of Medicine & Institute of Human Genetics, University of California, San Francisco, School of Medicine

**KIMBERLY J. POPOVITS,** President and CEO, Genomic Health, Inc.

**AIDAN POWER,** Vice President and Global Head of Molecular Medicine, Pfizer, Inc.

**RONALD PRZYGODZKI,** Associate Director for Genomic Medicine Biomedical Laboratory Research and Development, Department of Veterans Affairs

**AMELIE G. RAMIREZ** (IOM), Dielmann Chair, Health Disparities and Community Outreach Research and Director, Institute for Health Promotion Research, University of Texas Health Science Center at San Antonio

**LAURA LYMAN RODRIGUEZ,** Senior Advisor to the Director for Research Policy, National Human Genome Research Institute

**ALLEN D. ROSES** (IOM), Jefferson-Pilot Professor of Neurobiology and Genetics, Professor of Medicine, Director, Deane Drug Discovery Institute, and Senior Scholar, Fuqua School of Business R. David Thomas Executive Training Center, Duke University

**KEVIN A. SCHULMAN,** Professor of Medicine & Business Administration; Director, Center for Clinical and Genetic Economics; and Associate Director, Duke Clinical Research Institute, Duke University Medical Center

**SHARON TERRY,** President and CEO, Genetic Alliance

**STEVEN TEUTSCH,** Representative, Secretary's Advisory Committee on Genetics, Health and Society

**MARTHA TURNER**, Assistant Director, Center for Ethics and Human Rights, American Nurses Association

**MICHAEL S. WATSON,** Executive Director, American College of Medical Genetics

**CATHERINE A. WICKLUND,** Director, Graduate Program in Genetic Counseling; Immediate Past President, National Society of Genetic Counselors; and Assistant Professor, Department of Obstetrics and Gynecology, Northwestern University

**Staff**

**LYLA HERNANDEZ**, Staff Director (until March 2010)
**ADAM BERGER**, Project Director (from March 2010)
**ERIN HAMMERS**, Research Associate
**ALEX REPACE**, Senior Project Assistant

## NATIONAL CANCER POLICY FORUM

**HAROLD MOSES** (IOM), (Chair), Director Emeritus, Vanderbilt-Ingram Cancer Center

**FRED APPELBAUM**, Director, Clinical Research Division, Fred Hutchinson Cancer Research Center

**PETER BACH**, Associate Attending Physician, Memorial Sloan-Kettering Cancer Center

**EDWARD BENZ, JR.**, President, Dana Farber Cancer Institute

**THOMAS BURISH**, Past-Chair, American Cancer Society Board of Directors and Provost, Notre Dame University

**MICHAELE CHAMBLEE CHRISTIAN**, Former Director, Division of Cancer Treatment and Diagnosis, National Cancer Institute

**ROBERT ERWIN**, President, Marti Nelson Cancer Foundation

**BETTY FERRELL**, Research Scientist, City of Hope National Medical Center

**JOSEPH FRAUMENI, JR.**, Director, Division of Cancer Epidemiology and Genetics, National Cancer Institute

**PATRICIA GANZ**, Professor of Health Services, School of Public Health and Professor of Medicine, David Geffen School of Medicine, University of California, Los Angeles

**ROBERT GERMAN**, Associate Director for Science, Division of Cancer Prevention and Control, Centers for Disease Control and Prevention

**ROY HERBST**, Chief, Thoracic/Head & Neck Medical Oncology, M.D. Anderson Cancer Center

**THOMAS KEAN**, Executive Director, C-Change

**JOHN MENDELSOHN**, President, M.D. Anderson Cancer Center

**JOHN NIEDERHUBER**, Director, National Cancer Institute

**DAVID PARKINSON**, President and CEO, Oncology Research and Development, Nodality Inc.

**SCOTT RAMSEY**, Member, Cancer Prevention Program, Division of Public Health Science, Fred Hutchinson Cancer Research Center

**JOHN WAGNER**, Executive Director, Clinical Pharmacology, Merck & Company, Inc.

**JANET WOODCOCK**, Deputy Commissioner and Chief Medical Officer, U.S. Food and Drug Administration

### Staff

**ROGER HERDMAN**, Director (until September 2009)
**ADAM SCHIKEDANZ**, Staff (until June 2009)